LABORATORY TESTING
of
SOILS, ROCKS
and
AGGREGATES

Sivakugan | Arulrajah | Bo

J.ROSS
PUBLISHING

Copyright © 2011 by J. Ross Publishing

ISBN-13: 978-1-60427-047-1

Printed and bound in the U.S.A. Printed on acid-free paper

10 9 8 7 6 5 4 3 2 1

Library of Congress Cataloging-in-Publication Data

Sivakugan, N. (Nagaratnam), 1956-
    Laboratory testing of soils, rocks, and aggregates/by N. Sivakugan, A. Arulrajah,
and M.W. Bo
        p. cm.
    Includes bibliographical references and index.
    ISBN 978-1-60427-047-1 (pbk. : alk. paper)
    1. Soil mechanics—Laboratory manuals. 2. Soil surveys—Laboratory
manuals. I. Arulrajah, A., 1969- II. Bo, Myint Win, 1954- III. Title.
  TA710.S537 2011
  624.1'513—dc23
                            2011015499

This publication contains information obtained from authentic and highly regarded sources. Reprinted material is used with permission, and sources are indicated. Reasonable effort has been made to publish reliable data and information, but the author and the publisher cannot assume responsibility for the validity of all materials or for the consequences of their use.

All rights reserved. Neither this publication nor any part thereof may be reproduced, stored in a retrieval system, or transmitted in any form or by any means, electronic, mechanical, photocopying, recording or otherwise, without the prior written permission of the publisher.

The copyright owner's consent does not extend to copying for general distribution for promotion, for creating new works, or for resale. Specific permission must be obtained from J. Ross Publishing for such purposes.

Direct all inquiries to J. Ross Publishing, Inc., 5765 N. Andrews Way, Fort Lauderdale, FL 33309.

Phone: (954) 727-9333
Fax: (561) 892-0700
Web: www.jrosspub.com

To the laboratory technicians and technical officers, the unsung heroes

Contents

Test No.	Description	Page No.
	Part A: Introduction	

| | **Part B: Soil Testing** | |

Part C: Rock Testing

Part D: Aggregate Testing

Part E: References

Preface

The first soil testing book was written by Professor T.W. Lambe of MIT in 1951. The book covered 13 different soil tests in relatively good detail. This was widely used as a reference book for laboratory testing of soils for several decades. This 165-page classic is currently out of print. Since then, there have been a few soil testing books written by Bowles (1986), Das (2008), Day (2001), Germaine and Germaine (2009), Head (1992b, 1992c, 2006), and Liu and Evett (2009). The three-part manual by K.H. Head discusses the laboratory tests in great length, getting into the nitty-gritty details of each test and is quite valuable in troubleshooting. The second and third volumes closely follow the laboratory equipment manufactured by ELE International, one of the world's largest suppliers of soil testing equipment. For triaxial testing, *The Measurement of Soil Properties in the Triaxial Test* by A.W. Bishop and D.J. Henkel (1962) is probably the most popular book.

Lately, with the advancements in geotechnical and geo-environmental engineering, there have been substantial developments on the soil testing front. With the availability of new technology and our need to work with new materials (e.g., recycled aggregates, biosolids, mine tailings, and other alternative materials), new laboratory tests have been developed and old test procedures have been modified. These have resulted in the requirement for new laboratory test methods and revisions or updates of existing standards.

In geotechnical engineering, laboratory testing should not be limited to soils; there are rocks and aggregates. It is not always possible to avoid getting into rock mechanics, which is not covered in traditional geotechnical engineering education. Most smaller soil testing laboratories have no capacity to perform rock tests. Virgin and recycled aggregates of different sizes are being used in roadwork and often require tests that are not covered in most laboratory testing manuals.

This is the first book that covers the *laboratory tests of soils, rocks, and aggregates*. The objective of this book is to describe these tests in sufficient detail but in a concise manner without sacrificing the salient features of the tests. Sample data sheets are provided with all computations clearly explained. ASTM International (ASTM), originally known as the American Society for Testing and Materials, standards were followed wherever possible. For rock tests, the methods suggested by International Society for Rock Mechanics are followed closely and any deviations from ASTM are noted. Some of the aggregate tests are adopted from the British standards where they are covered more extensively. This book can be seen as a one-stop shop and the first point of reference for the main laboratory tests concerning soils, rocks, and aggregates. Irrespective of what is discussed in the book, the hardcore soil testers will still religiously adopt the specific standards they are required to follow.

Part A of the book covers the general introductory material on laboratory testing, equipment, measurements, and units. Parts B, C, and D cover the tests on soils, rocks, and aggregates, respectively. The authors have extensive experience in all phases of laboratory testing of soils, rocks, and aggregates, including an intimate knowledge of the laboratory equipment and the test procedures.

We are grateful to several people who have contributed to this book. Most importantly, Warren O'Donnell, the senior technical officer of the Soils and Rocks Laboratory at James Cook University, Townsville, critically reviewed most of the chapters of this book and made valuable suggestions. He also provided several photographs and datasheets included in this book.

N. Sivakugan, A. Arulrajah, and M.W. Bo

About the Authors

Dr. Nagaratnam Sivakugan is an associate professor and the head of Civil & Environmental Engineering in the School of Engineering and Physical Sciences at James Cook University, Townsville, Australia. He is a co-author of *Geotechnical Engineering: A Practical Problem Solving Approach*, a popular textbook adopted by many universities worldwide. He graduated from the University of Peradeniya, Sri Lanka, with first class honors and received his MSCE and PhD from Purdue University, Lafayette, Indiana. As a chartered professional engineer and registered professional engineer of Queensland, he does substantial consulting work, including extensive laboratory and in situ tests for geotechnical and mining companies throughout Australia as well as internationally. He is a Fellow of Engineers Australia. Dr. Sivakugan has supervised eight PhDs to completion and has published 65 scientific and technical papers in refereed international journals, 65 more in refereed international conference proceedings, and six book chapters. He serves on the editorial board of the *International Journal of Geotechnical Engineering* and is an active reviewer for more than 10 international journals. He developed a suite of geotechnical PowerPoint slideshows that is being used worldwide as an effective teaching and learning tool.

Dr. A. Arulrajah is an associate professor in geotechnical engineering at Swinburne University of Technology, Melbourne, Australia. He completed his BSc in civil engineering at Purdue University, Lafayette, Indiana, in 1992; MEngSc at the University of Malaya, Malaysia, in 2003 and PhD from Curtin University, Australia, in 2005. Dr. Arulrajah is a Fellow of Engineers Australia, and a Chartered Professional Engineer in Australia. Prior to his first academic appointment in 2006, he worked for 14 years for various engineering consultants in Australia, Singapore, and Malaysia. Dr. Arulrajah was the first author of a journal paper on ground improvement that won the Telford Premium Prize (2009) from the Institution of Civil Engineers, United Kingdom. He is also the recipient of the Shamsher Prakash prize for excellence in the practice of geotechnical engineering (2010) and the recipient of three Swinburne University Vice-Chancellor Awards. He is the author of nearly 40 scientific and technical papers in refereed international journals and 40 more in refereed international conference proceedings. He has supervised three PhD candidates to completion in these areas of research. He is an editorial board member of the *International Journal of Geotechnical Engineering* and is an active reviewer for several international journals.

Dr. Myint Win Bo (BO Myint Win) is a senior principal/director (Geo-Services) at DST Consulting Engineers, Canada. He graduated with a BSc (geology) from the University of Rangoon and received a postgraduate diploma in hydrogeology from University College, London and an M.Sc degree from the University of London. He obtained his PhD in civil engineering (specializing in geotechnics) from the Nanyang Technological University, Singapore. He is a Fellow of the Geological Society, London, and a Fellow of the Institution of Civil Engineers, UK. He is also a professional engineer, professional geoscientist, international professional engineer, United Kingdom, and a chartered geologist, scientist, engineer, environmentalist, European geologist, and European engineer. Dr. Bo is also serving in several national and international professional societies as a committee member. Dr. Bo is an experienced practicing engineer as well as educator, and he has been giving several special lectures and workshops at international conferences, tertiary institutions, and professional associations. Additionally, Dr. Bo is an adjunct professor at the University of Ottawa and Lakehead University in Canada as well as adjunct professor at the Swinburne University of Technology in Australia. He is the author of more than 50 scientific and technical papers in refereed international journals, 70 more in refereed international conference proceedings, and three textbooks. Dr. Bo is an active reviewer for more than 6 international journals.

This book has free material available for download from the
Web Added Value™ resource center at *www.jrosspub.com*

At J. Ross Publishing we are committed to providing today's professional with practical, hands-on tools that enhance the learning experience and give readers an opportunity to apply what they have learned. That is why we offer free ancillary materials available for download on this book and all participating Web Added Value™ publications. These online resources may include interactive versions of material that appears in the book or supplemental templates, worksheets, models, plans, case studies, proposals, spreadsheets, and assessment tools, among other things. Whenever you see the WAV™ symbol in any of our publications, it means bonus materials accompany the book and are available from the Web Added Value™ Download Resource Center at www.jrosspub.com.

Downloads for *Laboratory Testing of Soils, Rocks and Aggregates* consist of spreadsheets that can be used to develop laboratory specific datasheets for most tests and easily modified to your style and standards.

Introduction

Part A

Laboratory testing of soils, rocks, and aggregates is an integral part of the geotechnical design. The design parameters are derived from laboratory and in situ testing of the geomaterials. During the site investigation program, when the in situ tests (e.g., standard or cone penetration) are being carried out, it is a common practice to take samples from the ground at various locations for further laboratory tests.

The laboratory tests have certain advantages over in situ tests that include:

- Well defined boundary conditions that can also be controlled
- More rational interpretation
- Higher degree of accuracy in the measurements

On the other hand, in situ tests are quicker, test a larger volume of soil, and are relatively inexpensive. However, their boundary conditions are not well defined, and their interpretation is often empirical or semi-empirical. Does it make one better than the other? Not really. A good site investigation program will include both in situ and laboratory tests that complement each other. The use of one should not be at the expense of the other.

DISTURBED AND INTACT SAMPLES

The grains of a gravelly, sandy, or silty soil are equidimensional, where the dimensions in all three mutually perpendicular directions are of the same order of magnitude. In addition, they are nonplastic and noncohesive. Their packing density is measured by *relative density* (D_r) defined as:

$$D_r = \frac{e_{max} - e}{e_{max} - e_{max}} \times 100 \tag{A.1}$$

where e_{max} = void ratio at the loosest state and e_{min} = void ratio at the densest state (e = current void ratio at which the relative density is computed). D_r varies between 0 and 100%. At the loosest state, it is 0 and at the densest state it is 100%. Classification of granular soils based on packing density of the grains is shown in Figure A.1. The behavior of a granular soil subjected to external loading is highly dependent on the grain size distribution, relative density, and the angularity of the soil grains.

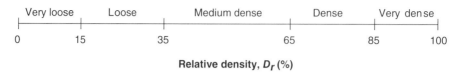

Figure A.1 Classification of granular soils based on relative density

Clay particles are one- or two-dimensional with shapes of needles or flakes and have a large *specific surface* (i.e., surface area per unit mass, measured in m²/g). For example, *montmorillonite* clays can have a specific surface of 800 m²/g. They are electrically charged with a net negative charge. In the presence of water, they are sticky and can exhibit plasticity and cohesion. Clayey soils can be classified on the basis of their plasticity and unconfined compressive strength as shown in Figures A.2 and A.3, respectively. Undrained shear strength (c_u) is half of the unconfined compressive strength (q_u). The term relative density is not applicable to clays.

Undisturbed samples literally means that the samples have the same characteristics (e.g., temperature and stresses) as in the in situ state. It is impossible to remove material from the ground without causing a disturbance to at least one of these characteristics. *Intact samples*, on the other hand, are the ones obtained using the best possible methods to minimize disturbance and still acknowledge that there is a degree of disturbance. The terms *samples* and *specimens* are misused commonly in practice. Samples are what we collect from the site, and from these samples we select specimens for specific laboratory tests. Specimen is a subset of sample.

It is difficult to collect good quality intact or undisturbed samples in granular soils. If required, there are some special techniques such as ground freezing, using resins, or attaching a core catcher to the sampling tube, among others. When laboratory tests are required on granular soils, it is common to perform them on *reconstituted* samples. Here, the granular soil grains

Figure A.2 Classification of clays based on plasticity (after Burmister 1949)

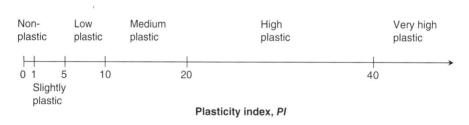

Figure A.3 Classification of clays based on unconfined compressive strength

are repacked to densities that are representative of the field samples and hence replicate the field situation. Reconstituted cohesive soil specimens can be prepared in the *Harvard miniature compaction device* (Wilson 1970). This device provides a kneading action on the soil placed in layers and simulates compaction by a sheepsfoot roller. It can be used in the laboratory to produce specimens that can be tested directly in a triaxial or uniaxial compression setup without further preparation. Reconstituted clay samples can also be prepared by sedimenting the clay slurry, typically mixed at a water content of 1.5 to 2.5 times the liquid limit.

In cohesive soils, good quality intact samples are required for laboratory tests such as triaxial, direct shear, consolidation, and permeability tests. They can be obtained from boreholes or trial pits. The samples from boreholes are recovered from thin walled *Shelby tubes*™ or special samplers such as a *piston sampler*. Piston samplers are effective in very soft clays and organic soils such as peat. Hvorslev (1949) introduced area ratio A_R as an indirect measure of the degree of mechanical disturbance that can be expected. It is defined as:

$$A_R(\%) = \frac{D_o^2 - D_i^2}{D_i^2} \times 100 \qquad (A.2)$$

where D_i and D_o are the inner and outer diameter of the sampler. For a good quality intact sample, A_R should be less than 10%. In very stiff or hard strata, it may be necessary to use tubes with greater wall thicknesses and, thus, significantly higher values of A_R. It is common to see *double* or *triple tube core barrel samplers* used in very stiff or hard soils for obtaining good quality intact samples. Hvorslev suggested that the length of the intact sample be limited to 10 to 20 times the diameter of the tube in cohesive soils. ASTM International (ASTM) D1587 suggests the maximum length to be 910 mm (36 in) for 50 to 75 mm (2 to 3 in) diameter tubes, and 1450 mm (57 in) for 127 mm (5.0 in) diameter tubes. The sampling tubes are made of mild, galvanized, or stainless steel, or brass. When sampling in environmental projects, to avoid chemical reactions with the metal, epoxy-coated steel or plastic liners can be used.

Sample disturbance occurs due to two separate factors, namely, *mechanical disturbance* and *stress-relief*. Mechanical disturbance is caused by the drilling equipment and the sampler that is inserted into the borehole. As seen in Equation A.2, the mechanical disturbance increases with the wall thickness. As a result, the outer annular region in the sample recovered from the borehole can be disturbed. Under such circumstances, it is a good practice to discard the annular region and use the core from the center. To overcome the effects of stress relief, the sample can be reconsolidated to the estimated in situ overburden stress. Block samples can be recovered from a wall or base of an excavation or trial pit.

Disturbed or remolded samples are adequate for the soil classification, Atterberg limits, water content, and specific gravity determination. They can also be used for compaction and California Bearing Ratio tests if available in sufficient quantity. The split-spoon sampler used in a standard penetration test has A_R in excess of 100%, and hence the samples are highly disturbed, rendering them suitable only for visual identification purposes. The split-spoon sampler can be equipped with a liner that will hold the samples intact. The liner can then be sealed and waxed at the ends before it is transported to the laboratory to preserve the natural water content and avoid oxidation. Sealable plastic bags and glass jars with air-tight lids are useful for preserving the samples at their natural water content. When filled with samples, they have to be properly labeled indicating the project, geographic location, sample depth, date, and other relevant information.

For safety reasons, it is a common practice to fill the borehole or trial pit with soil once the sampling and the associated in situ tests are completed.

ACCURACY, PRECISION, AND RESOLUTION

Let's define some simple terms associated with laboratory measurements. *Resolution* is the smallest change the measuring device can display (e.g., 0.01 g in a digital balance). In a digital display, the resolution is simply one digit of change in the last digit. The term *accuracy* can be split into two components: *precision* and *bias*. Precision is a measure of scatter about an average value of the measurements that need not be close to the true value. The difference between this average of the measured values and the true value is known as the bias of the instrument. *Sensitivity* refers to the response of a device to a unit input (e.g., 100 milli volts per mm movement of a linear variable differential transformer). Sensitivity applies to measuring devices such as transducers, load cells, or dial gages; resolution applies to the readout or display devices. *Repeatability* is slightly different from *reproducibility*. Repeatability is a qualitative measure of the variability between test results when the tests are repeated on identical specimens by the same person in the same laboratory under the same conditions. Reproducibility is a measure of variability between the test results when the tests are repeated in different laboratories, by different operators using different equipment.

The test measurements and computed results should be reported to appropriate *significant digits*. When reporting the mass of a soil sample as 142.3 g, there are four significant digits. Reporting too many digits can give a false sense of accuracy and demonstrates a lack of appreciation for the quality of measurements. The significant digits should be based on the sensitivity of the instrument, resolution of the measuring device, specimen size, and measured quantity. In geotechnical laboratory testing, it is sometimes specified by the relevant standards. For example, liquid or plastic limit and plasticity index are generally rounded to the nearest integers. Densities (in t/m^3 or Mg/m^3) and specific gravities are reported to 0.01.

SOME LABORATORY DEVICES

In this section, we discuss some common laboratory devices that are used frequently in the laboratory testing of soils, rocks, and aggregates.

Length Measuring Devices

Length measurement devices used in a geotechnical laboratory include a meter stick, ruler, measuring tape, Vernier calipers, micrometers, dial gages, and LVDTs, among others. Some of these are shown in Figure A.4. Vernier calipers and micrometers are used for precise one-off measurements of dimensions such as inner and/or outer diameters of a cylinder. Dial gages and LVDTs are used for precise continuous measurements of displacements or deformations during a test.

Mass Measuring Devices

Water content is one of the most common measurements in the geotechnical laboratory. It is carried out as a part of most laboratory tests and when collecting field samples on samples as

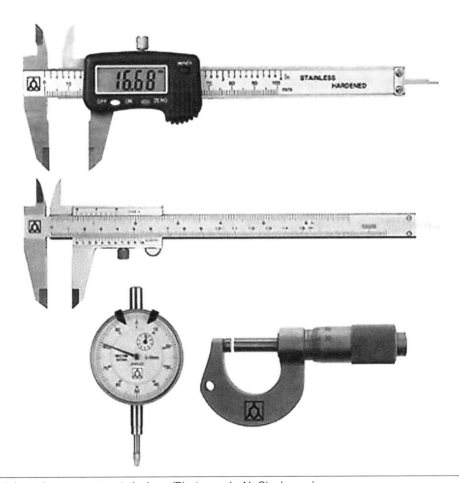

Figure A.4 Length measurement devices (Photograph: N. Sivakugan)

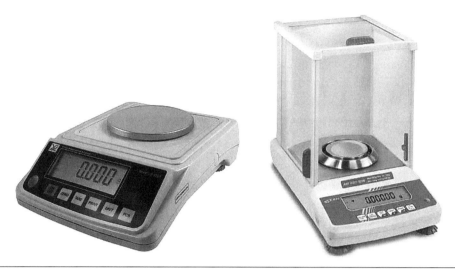

Figure A.5 Mass measurement devices (Photograph: N. Sivakugan)

small as 20 g in mass. A good balance (Figure A.5) capable of measuring the mass to 0.01 g resolution is preferable. This includes top pan balances and analytical balances. When dealing with larger soil samples, it may be necessary to use a coarse balance such as a semi-self-indicating scale or a heavy platform scale with lesser resolution.

Load Measuring Devices

Frequently, there is a requirement to measure the load acting on a specimen rather precisely. It is necessary to monitor the loads throughout the tests, especially in strength tests where the specimens are tested to failure. *Load cells* and *proving rings* are some of the common devices that are used for this purpose (Figure A.6). Pressures are generally measured by transducers.

Drying Equipment

To determine water contents and to dry soils in preparation for laboratory tests, it is necessary to dry out the soils completely. Laboratory drying ovens are useful for this purpose. They are usually set at 105 to 110°C for inorganic soils and somewhat lower for organic soils. Drying for 16 to 24 hours is usually sufficient. By plotting the dry mass with time, this can be verified. When the difference between successive weighings is less than 0.1% of the dry mass, it can be assumed that the soil has reached a constant mass. Metal trays, water content tins, and weighing bottles are suitable for drying soils in the oven. They should be transferred to the *desiccator* and cooled without absorbing any moisture before placing on a weighing scale or balance.

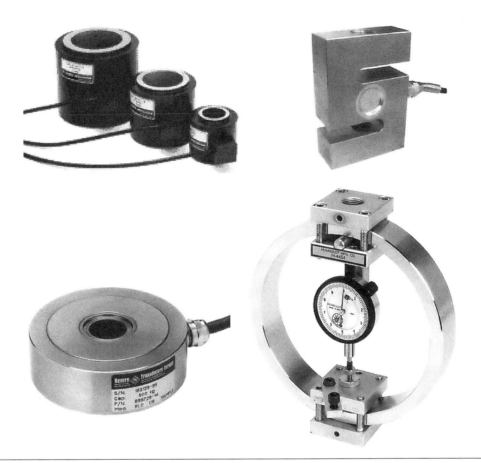

Figure A.6 Load cells and proving ring (Photograph: N. Sivakugan)

Desiccator

A desiccator (Figure A.7) is a glass enclosure that contains a desiccant (e.g., anhydrous silica gel) that keeps the air within the desiccator dry. It is used to cool a sample without absorbing moisture when removed from the oven. With its airtight seal, it can be used to contain the samples when applying vacuum. The silica gel, in the form of crystals, can be spread in a dish and kept below the perforated floor of the desiccator. The self-indicating crystals are blue in color when dry and slowly become colorless when they absorb moisture. By heating them in the oven at 110°C, they can be dried and reused. There are also cheaper nonindicating desiccants that do not change color. Desiccators can be useful for containing the soil specimens while de-airing through application of vacuum.

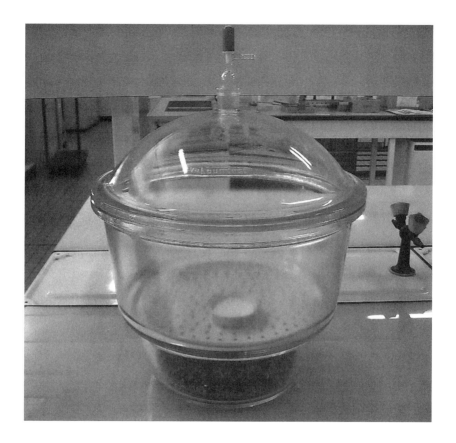

Figure A.7 Desiccator (Photograph: N. Sivakugan)

Constant Temperature Baths

For laboratory tests (e.g., hydrometer tests) that require a constant temperature environment for a long period of time, a constant temperature bath is useful.

Vacuum Pump

Often the laboratory tests are carried out on saturated soil specimens. De-airing water, soils, and soil-water mixes can be accomplished by applying vacuum and/or by heating. Remember that atmospheric pressure at sea level is 1 atm (= 760 mm Hg = 101.325 kPa = 1.01325 bar). Therefore, theoretically, the maximum vacuum one can apply is 101.325 kPa. A vacuum pump with a pressure gage for measuring the vacuum is a useful device in a geotechnical laboratory. Vacuum is commonly applied to specimens placed in a desiccator that can be covered by a protective cage for safety reasons.

Water Purification System

Tap water contains dissolved ions and bacteria that can react with the soil samples that are being tested. Therefore, it is necessary to use some purer forms of water such as *distilled* or *deionized water* for the laboratory tests. Distilled water has all of the impurities removed by boiling the water and condensing the steam. Deionized water has all of the mineral ions removed by running the water through a series of filters. In most soil testing systems, the tests are carried out on saturated specimens. To ensure no air is introduced into the system, *de-aired* distilled or deionized water is used. De-aired water does not have any dissolved air in it. De-airing can be carried out by applying vacuum to the water in a suitable vessel that can resist the atmospheric pressure from outside.

Wax Pot And Wax

Wax is effectively used for sealing samples in tubes or other containers, protecting them against any moisture changes until the tests are carried out. To avoid damage to the samples, wax with a low melting point is preferable. Wax with a high melting point tends to be more brittle and, hence, is prone to cracking. Paraffin is the most common wax that is used for sealing sampling tubes. It is inexpensive and is available from hardware stores. Being relatively brittle, this can crack after a few days and destroy the seal. Germaine and Germaine (2009) and Lambe (1981) recommend mixing paraffin with petroleum jelly in a 1:1 ratio. An electrically heated wax pot with thermostat control can maintain the molten wax at the right temperature while coating the samples and sealing tubes. A wire brush to apply the wax and a ladle for collecting wax from the wax pot are other items that can help in the use of a wax pot and wax. Samples can also be waxed by dipping them in the molten wax. Wax should not be heated to more than a few degrees above its melting point. Overheating can affect the sealing properties and the wax can become brittle.

STANDARDS

The laboratory and field tests are generally conducted according to some standards that are developed by specialists in the area. Some of these specialists are ASTM Intermational (ASTM), International Standardisation Office (ISO), American Association of State Highway and Transportation Officials (AASHTO), Australian Standards (AS), and British Standards (BS).

ASTM standards (http://www.astm.org) cover a wide range of materials, including steel, concrete, soils, rocks, petroleum, paints, and water. The standards are published in more than 80 volumes. Luckily, all that concerns us are available in two volumes, each of which has more than 1000 pages. The volumes are also available separately on CDs. ASTM standards for soils, aggregates, and rocks are updated regularly by Committee D18 and are published in two volumes 04.08 and 04.09 that come under Section 4—Construction.

UNITS

Worldwide, two of the most commonly used systems of units are the *Imperial* (British) and *metric* systems. In addition to these two systems, there have been a few hybrids where the units were simply derived from these two. Since the 1950s, in an attempt to standardize the units across the globe, SI units have become increasingly popular. SI stands for "Le Système International d'Unités" (The International System of Units). It is the modern form of the metric system. The United Kingdom converted to SI in 1972, followed by Australia and New Zealand. In North America, both systems are still being used.

The SI system is based on the following seven base units:

- Length in meter (m)
- Mass in kilogram (kg)
- Time in second (s)
- Electric current in ampere (A)
- Thermodynamic temperature in Kelvin (K)
- Luminous intensity in candela (cd)
- Amount of substance in mole (mol)

Only the first three are commonly used in geotechnical engineering. The units for all other physical quantities are derived from all of the seven basic units. Some of the derived quantities have been given specific names. For example:

- Newton (N) = Force required to accelerate a mass of 1.0 kg at the rate of 1.0 m/s^2
- Pascal (Pa) = N/m^2
- joule (J) = Work done when a force of 1 Newton moves a distance of 1 m.
- watt (W) = joule/s

Other special names, not necessarily SI but worth noting are:

- Ångström (Å) = 10^{-10} m
- bar = 10^5 Pa
- British (short) ton = 2000 lb
- British (long) ton = 2240 lb
- dyne = 10^{-5} N
- hectare (ha) = 10000 m^2
- hertz = 1.00 cycle/s
- hundred weight (UK) = 112 lb
- hundred weight (US) = 100 lb
- kilopond = kilogram-force
- kip = 1000 lb-force (lbf)
- metric ton (tonne) = 1000 kg
- nautical mile = 1852 m
- poundal = Force required to accelerate a mass of 1.0 lb at the rate of 1.0 ft/s^2
- stone (US) = 12.5 lb; stone (UK) = 14.0 lb

Some unit conversions that are commonly used in geotechnical engineering are:

Length:

- angström = 10^{-10} m
- foot (ft) = 0.3048 m
- inch = 2.5400×10^{-2} m
- mil = 1/1000 inch = 2.54×10^{-5} m
- mile = 1.609344 km
- nautical mile = 1852 m
- yard = 3.00 ft = 0.9144 m

Area:

- acre = 4840 sq. yard = 4046.8564224 m^2
- hectare (ha) = 10,000 m^2
- perch = 25.29285264 m^2
- square = 9.290304 m^2

Volume:

- gallon (U.S.) = 3.785412×10^{-3} m^3
- gallon (U.K.) = 4.546090×10^{-3} m^3
- liter = 1.0×10^{-3} m^3

- ounce (U.S. fluid) = 2.957353×10^{-5} m^3
- pint (U.S. fluid) = 4.731765×10^{-4} m^3

Mass:

- British ton (short) = 2000 lb
- British ton (long) = 2240 lb = 20 hundredweight
- hundredweight (U.K.) = 112 lb
- hundredweight (U.S.) = 100 lb
- metric ton (or tonne) = 1000 kg
- ounce (mass) = 28.34952 g
- pound (mass) = 0.4535924 kg
- U.S. ton = 2000 lb

Force:

- dyne (dyn) = 1.00×10^{-5} N
- kilogram-force = 1.00 kilopond = 9.80665 N
- kip-force (1000 lbf) = 4448.222 kN
- poundal = 0.13825495 N
- pound force (lbf) = 4.448222 N

Pressure:

- atmosphere (760 mm Hg) = 1.013247×10^5 Pa
- bar = 1.0×10^5 Pa
- kgf/cm^2 = 98.06650 kPa
- lbf/in^2 (psi) = 6.894757 kPa
- kips/in^2 (ksi) = 6.894757 MPa
- ton-force (short)/sq. ft = 95.760518 kPa
- ton-force (long)/sq. ft = 107.251780 kPa

Work or Energy:

- Btu = 1055.0559 J
- calorie = 4.186800 J
- erg = 1.00×10^{-7} J
- joule (J) = 1.000 N(m
- kilowatt-hour = 3600 kJ

Dynamic viscosity:

- poise (P) = 0.1 Pa·s
- centipoise (cP) = 0.001 Pa·s

Kinematic viscosity (dynamic viscosity divided by density):

- Stoke (St) = 1.0 cm^2/s = 1.0 × 10^{-4} m^2/s
- Centistoke (cSt) = 1.0 × 10^{-6} m^2/s

Prefixes indicate orders of magnitude in steps of 1000 and provide a convenient way to express small and large numbers. The prefixes used with SI units are summarized in Table A.1. Some notes from ASTM that should be observed when expressing SI units are:

a. In derived units formed by multiplication, use *raised* dots (·) between the symbols. For example; N·m or Pa·s, for example. The unit names should be written as Newton meter, Pascal second, and so forth with space between the different unit names.
b. In derived units formed by division, use only one solidus (/) per expression and parentheses to avoid ambiguity.
c. SI symbols are not abbreviations and hence no period should follow them except at the end of a sentence.
d. Unit is the same whether singular or plural.

Table A.1 SI prefixes used with units

Factor	Prefix	Symbol
10^{24}	yotta	Y
10^{21}	zeta	Z
10^{18}	exa	E
10^{15}	peta	P
10^{12}	tera	T
10^{9}	giga	G
10^{6}	mega	M
10^{3}	kilo	k
10^{2}	hecto	h
10^{1}	deca	da
10^{-1}	deci	d
10^{-2}	centi	c
10^{-3}	milli	m
10^{-6}	micro	μ
10^{-9}	nano	n
10^{-12}	pico	p
10^{-15}	femto	f
10^{-18}	atto	a
10^{-21}	zepto	z
10^{-18}	yocto	y

e. Leave a space between the value and the symbol. For example, 5 m, not 5m. An exception is for the plane angle degree which is expressed as 73° and not 73°. Symbol for degree Celsius is °C.

f. Do not place a space or hyphen between the prefix and the unit name. For example, kilogram, not kilo-gram.

g. Choose a prefix so that the number lies between 0.1 and 1000.

LABORATORY REPORT

All the salient findings from a laboratory testing program must be documented clearly in a laboratory report that is submitted to the client. The information presented should be factual, nonambiguous, and should include:

- Name and address of the laboratory
- Name of the client, project, and sample location
- Visual classification and description of the test samples
- Standards followed and deviations in the procedure (if any)
- Properly completed datasheets
- Analysis and interpretation of the test results
- Name and signature of the tester with date

Use of spreadsheets can streamline the calculations involved in the test. Most commercial laboratories would prefer using their standardized spreadsheet for this purpose. Table A.2 lists the Greek symbols that are used to denote some variables in engineering in general.

SAFETY

These days we are realizing the value of workplace health and safety more and more. Laboratories with machinery, equipment, and hazardous material are some of the high risk areas prone to accidents. Material Safety Datasheets must be used to assess whether special protective clothing and/or eye protection are required. Hazardous chemicals should be disposed of in an appropriate manner. We must take every possible step to minimize the risk. There are designated workplace health and safety officers to give induction for the beginners and ensure that those working in the laboratories follow the best practices as far as the workplace health and safety is concerned.

FURTHER READING

The late Professor T. W. Lambe of Massachusetts Institute of Technology wrote the first reference book *Soil Testing for Engineers* in 1951. There were no further updated editions of this book,

Table A.2 Greek alphabet

Name	Symbol		Name	Symbol	
	Uppercase	**Lowercase**		**Uppercase**	**Lowercase**
alpha	A	α	nu	N	ν
beta	B	β	xi	Ξ	ξ
gamma	Γ	γ	omicron	O	o
delta	Δ	δ	pi	Π	π
epsilon	E	ε	rho	P	ρ
zeta	Z	ζ	sigma	Σ	σ
eta	H	η	tau	T	τ
theta	Θ	θ	upsilon	Υ	υ
iota	I	ι	phi	Φ	φ
kappa	K	κ	chi	X	χ
lambda	Λ	λ	psi	Ψ	ψ
mu	M	μ	omega	Ω	ω

and it is currently out of print. This 165-page book has remained a valuable source of reference in many geotechnical laboratories worldwide. The U.S. Army Corps of Engineers produces numerous design manuals and guidelines that are available free and can be downloaded from their website http:///www.usace.army.mil/publications/eng-manuals. This includes *Laboratory Soil Testing* (USACE 1986). The U.S. Naval Facilities Engineering Command produces excellent design manuals that are widely used in geotechnical engineering. The three most commonly used manuals are: *DM7.01-Soil Mechanics*, *DM7.02-Foundations and Earth Structures*, and *DM7.03-Soil Dynamics and Special Design Aspects* downloaded from their website http://www.navfac.navy.mil. The U.S. Bureau of Reclamation produces *Earth Manual* which covers earthwork and the associated lab and in situ testing and construction control. This can be downloaded free from http://www.usbr.gov/pmts/writing/earth/earth.pdf. The American State Highway Transportation Officials (AASHTO) produces their standards in *Standard Specifications for Transportation and Methods of Sampling and Testing*. This can be purchased from AASHTO. The three volume series by K.H. Head, *Manual of Soil Laboratory Testing*, can be a useful reference source in a soil testing laboratory. They are quite comprehensive and cover the nuts and bolts of the test procedures and equipment.

This book has free material available for download from the
Web Added Value™ resource center at *www.jrosspub.com*

Soil Testing — Part B

INTRODUCTION

Soil testing is an integral part of a site investigation program. The disturbed and intact soil samples collected at the site are taken to geotechnical laboratories where specific tests are performed. Some tests (e.g., consolidation or triaxial tests) require good quality intact samples whereas disturbed samples are adequate for the others (e.g., water content or Atterberg limits).

This section of the book covers most of the important soil tests that are carried out in geotechnical engineering laboratories. This includes visual classification and identification of soils, grain size distribution through sieve and hydrometer analyses, Atterberg limits, compaction, permeability, consolidation, strength, and others. Rock tests are covered in Part C and aggregates in Part D. Some of the tests that are covered in Part D, for example, aggregate testing, are also applicable to soils. The soil tests described in this section can be broadly categorized as:

I. Soil classification
   a. Visual identification and classification of coarse grained soils
   b. Visual identification and classification of fine grained soils
   c. Water content
   d. Specific gravity
   e. Sieve analysis
   f. Hydrometer analysis
   g. pH
   h. Organic content
   i. Liquid limit by Casagrande's percussion cup
   j. Liquid limit by fall cone
   k. Plastic limit
   l Linear shrinkage
II. Earthwork
   a. Compaction
   b. Maximum dry density of a cohesionless soil
   c. Minimum dry density of a cohesionless soil
   d. Field density measurement
   e. California Bearing Ratio

III. Permeability
   a. Permeability of a coarse grained soil—constant head
   b. Permeability of a fine grained soil—falling head
IV. Consolidation
   a. One-dimensional consolidation by incremental loading
 V Strength
   a. Direct shear
   b. Consolidated, undrained triaxial
   c. Unconsolidated, undrained triaxial
   d. Unconfined, compressive strength

STANDARDS

The test procedures described here are based on the ASTM International standards (ASTM), the Australian standards (AS), and British standards (BS). The relevant standards are shown at the beginning of each test with an asterisk to signify the one that was followed more closely when it is not ASTM. For example, some of the aggregate tests are based on the British standards and most of the rock tests are based on the International Society of Rock Mechanics suggested procedures.

ASTM produces standards, in more than 75 volumes, for soils, rocks, aggregates, and other materials, including paint, textiles, rubber, and plastics. ASTM standards are published under several *sections*, one of which is "Section 4—Construction" which is subdivided into 13 volumes. All the standards relating to soils and rocks are covered in two volumes:

- *Volume 04.08 Soil and Rock (I): D420—D5876*
- *Volume 04.09 Soil and Rock (II): D5877—Latest*

The standards for aggregates are covered in *Volume 04.02 Concrete and Aggregates*. Geosynthetics are covered separately in *Volume 04.13 Geosynthetics*.

The Australian Standards cover material testing, designs, and construction. Some of the relevant standards are:

- AS 1289 Methods of testing soils for engineering purposes
- AS 1726 Geotechnical site investigations
- AS 2159 Piling—design and installation
- AS 2870 Residential slab and footing—construction
- AS 3798 Guidelines for earthworks on commercial and residential developments
- AS 4678 Earth-retaining structures
- AS 4969 Analysis of acid sulphate soils
- AS 5100.3 Bridge design—foundations and soil supporting structures

The British standards for soils tests are from BS1377:1990 which has been through a significant improvement from the previous version BS1377:1975 that was limited to a single document, little more than 100 pages. BS1377:1990 consists of nine parts. They are:

- Part 1: General requirements and sample preparation
- Part 2: Classification tests
- Part 3: Chemical and electrochemical tests
- Part 4: Compaction related tests
- Part 5: Compressibility, permeability and durability tests
- Part 6: Consolidation and permeability tests in hydraulic cells and with pore pressure measurement
- Shear strength tests (total stress)
- Shear strength tests (effective stress)
- In situ tests

The aggregates are covered separately in BS 812.

B1   Visual Identification and Classification of Coarse Grained Soils

Objective: To visually identify and classify a coarse grained soil

Standards: ASTM D2487 & D2488

Introduction

The person at the site classifying the samples is not the one who will do the designs and analysis in the office. Therefore, it is necessary to communicate the soil description as precisely as possible because the design office could be hundreds of kilometers away. A *soil classification system* becomes useful in this situation. It is a systematic method to group soils of similar behavior, and then to describe and classify them. The strict guidelines and the standard terms proposed within the system eliminate any ambiguity and make it a universal language among geotechnical engineers. There are several soil classification systems currently in use. The *Unified Soil Classification System* (USCS) (ASTM D2487) is the most popular one used in geotechnical engineering practice worldwide. The AASHTO classification system is quite popular for roadwork where soils are grouped according to their suitability as subgrade, embankment, sub-base, or base materials. There are also country-specific standards such as *Australian Standards* (AS), *British Standards* (BS), or *Indian Standards* (IS).

A good geotechnical engineer must be able to identify and classify soils in the field simply by the feel and appearance. This is easier with coarse grained soils where one can include qualitative information on *grain size* (e.g., fine, medium, or coarse), *grain shape, color, homogeneity, gradation, state of compaction* or *cementation, presence of fines*, and so forth. Based on relative proportions and this information, it is possible to assign the USCS symbol and a coherent description (ASTM D2488).

Soils can behave quite differently depending on their geotechnical characteristics. In coarse grained soils, where the grains are larger than 0.075 mm (75 μm), the engineering behavior is influenced mainly by the relative proportions of the different grain sizes present within the soil, density of their packing, and shapes of the grains. In fine grained soils, where the grains are smaller than 0.075 mm, the mineralogy of the soil grains and the water content have much greater influence than the grain sizes on the engineering behavior. The borderline between coarse and fine grained soils is 0.075 mm, which is the smallest grain size one can distinguish with the naked eye. Based on the grain sizes, soils can be grouped as clays, silts, sands, gravels, cobbles, and boulders as shown in Figure B1.1. It shows the borderline values as per the USCS, BS, and the AS. Within these major groups, soils can still behave quite differently, and we will look at some systematic methods of classifying them into distinct subgroups.

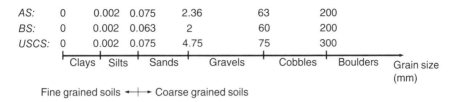

Figure B1.1 Major soil groups based on grain sizes

Procedure:

1. Select a representative sample.
2. Remove boulders and cobbles and estimate their percentage by weight.
3. Spread the remaining soil on a flat surface and check whether the coarse or fine grained soils are in majority. Remember, one can see the individual grains by the naked eye only in coarse grained soils. If the majority is coarse, classify the soil as a coarse grained soil (gravel or sand). If the majority is fine, classify the soil as a fine grained soil (silt or clay).
4. State whether the coarse grained soil is *clean* (i.e., contains less than 5% fines) or it has appreciable fines.
5. In clean coarse grained soil, state whether it is well graded or poorly graded. If there are appreciable fines, state whether the fines are silty or clayey. Use terms like *trace* (less than 5%), *few* (5 to 10%), *little* (15 to 25%), *some* (30 to 45%) and *mostly* (more than 50%) to describe the presence of grains of specific groups. Clays feel sticky and silts feel gritty when wet.
6. Note any of the following data wherever possible:
   ○ Grain size (at least qualitatively, as fine, medium, or coarse sand)
   ○ Grain angularity (e.g., angular, subangular, subrounded, or rounded) (see Figure B1.2)

Figure B1.2 Grain angularity

Regular: $w/t < 3$ and $l/w < 3$
Flat: $w/t > 3$
Elongated: $l/w > 3$

Figure B1.3 Grain shape

- Grain shape (e.g., regular, flat, elongated, or flat and elongated—based on three as the aspect ratio limit) (see Figure B1.3)
  - Moisture condition as dry, moist (damp), or wet (visible free water)
  - Color in conjunction with the moisture condition
  - Odor (only if organic or unusual)
  - Cementation between grains as weak, moderate, or strong
  - Reaction (formation of bubbles) to hydrogen chloride (HCl) to detect the presence of carbonates; describe as none, weak, or strong
7. USCS (or other) symbol and description of the soil based on the above data

ASTM suggests borderline symbols for soils with two possible identifications where the two symbols are separated by a solidus (e.g., GW/GP). The group name for such borderline soils should be the group name of the first symbol.

The borderline symbol applies to the following situations where some examples are also given:

- Soils containing approximately the same amount (45 to 55%) of coarse and fine grained soils: GM/ML or ML/GM.
- Coarse grained soils containing approximately the same amount of sands and gravels: GC/SC, GP/SP.
- Coarse grain soil that can be classified as well or poorly graded: GW/GP, SW/SP.

Here, G = gravel, S = sand, M = silt, C = clay, W = well graded, and P = poorly graded. Every effort must be made to avoid a borderline symbol; it should be used only if absolutely necessary.

Cost: US$10

B2 Visual Identification and Classification of Fine Grained Soils

Objective: To visually identify and classify a fine grained soil

Standards: ASTM D2487 & D2488

Introduction

Fine grained soils consist of silts and clays. The grains are smaller than 0.075 mm and are not visible to the naked eye. The grain size distribution of a fine grained soil is of little value. The purpose of this exercise is to identify the fines as one of the two broader groups, namely, silts and clays, and to further classify them on the basis of the plasticity that they display. The fines are identified on the basis of *dry strength*, *dilatancy*, *toughness*, and *plasticity*.

Dry strength: How easy is it to crush a dry lump by squeezing between the fingers? Describe dry strength as none, low, medium, high, or very high. Silts have low dry strength and clays have high dry strength.

Dilatancy: If a moist pat of fines is placed on the palm and shaken, how quick does the water appear on the surface and disappear upon squeezing? Use terms such as none, slow, or rapid. Silts show rapid dilatancy and clays show none to slow dilatancy.

Toughness: Plastic limit is the lowest water content at which the fines can be rolled into a 3 mm diameter thread. Toughness is a measure of the strength near the plastic limit. By kneading the moist pat of fines at a water content close to the plastic limit, it is possible to describe toughness as low, medium, or high. High plastic clays will have high toughness, and silts will have low toughness.

Plasticity: On the basis of the observation made while testing for toughness, describe plasticity as nonplastic, low, medium, or high. Silts are generally nonplastic and clays can have high plasticity.

An approximate guide to assign the Unified Soil Classification System (USCS) symbols based on these four is given in Figure B2.1.

Procedure:

1. Select a representative sample.
2. Use only the fraction passing a No. 40 (0.425 mm) sieve.
3. Test separately for dry strength, dilatancy, toughness, and plasticity. Use Figure B2.1 as the basis for deciding whether the fines are clay or silts and to assign a symbol. In addition, a moist pat of clay feels sticky and silt feels gritty.

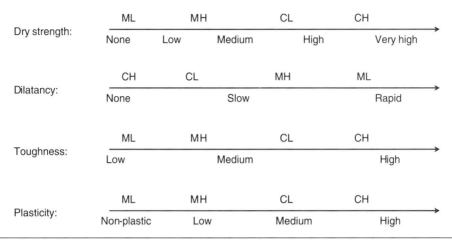

Figure B2.1 Visual classification of fine grained soils

4. Note any of the following data wherever possible:
    o Moisture condition as dry, moist (damp), or wet (with visible water)
    o Consistency (if moist or wet) as very soft, soft, firm, hard, or very hard
    o Color in conjunction with the moisture condition
    o Odor, especially for organic soils
    o Cementation between grains as weak, moderate, or strong
    o Reaction (formation of bubbles) to HCl for detecting the presence of carbonates; describe as none, weak, or strong.
5. USCS (or other) symbol and description of the soil, based on this data

While visually identifying and describing soils, sometimes it is quite likely that a soil can be assigned to one of the two groups. ASTM suggests borderline symbols for soils with two possible identifications, where the two symbols are separated by a solidus (e.g., CL/CH, GW/GP). *The group name for such borderline soils should be the group name of the first symbol*, with the following exceptions, where the group names are as follows:

- CL/CH lean to fat clay
- ML/CL silty clay
- CL/ML clayey silt

The borderline symbol applies to the following situations where some examples are also given.

- Soils containing approximately the same amount (45 to 55%) of coarse and fine grained soils: GM/ML or ML/GM; GP/SP; SC/GC, etc. ASTM D2488 notes that it is practically impossible to have GW/SW.

- A coarse grained soil that can be either well graded or poorly graded: GW/GP or SW/SP
- Fines that can be classified as clay or silt: CL/ML, CH/MH, SC/SM
- Fine grained soil that can be classified as having low or high compressibility: CL/CH, MH/ML

Here, G = gravel, S = sand, M = silt, C = clay, L = low plasticity, and H = high plasticity. Every effort must be made to avoid a borderline symbol; it should be used only if necessary.

Cost: US$10

B3   Water Content

Objective: To determine the water (moisture) content of a soil or aggregate sample by *oven drying*

Standards: ASTM D2216
AS 1289.2.1.1
BS 1377-2

Introduction

Water content determination is fairly straightforward and becomes a part of most laboratory tests. The most common method involves heating the sample in the drying oven for 24 hours. There are other methods using a microwave oven (ASTM D4643, AS 1289.2.1.4), infrared lights, and hot plate, which are not discussed here. Water content can vary from 0% for a dry soil to more than 1000% for slurries, organic, and soft soils. Water content measurements should be carried out on samples from the site at the earliest possible time before any moisture losses, corrosion of the sampling tube, and oxidation of the samples occur. In marine sediments containing significant salt content, special care is required to remove the salt or account for its presence in determining the water content.

Procedure:

1. Determine the mass m_1 of a clean and dry container. Metal cans made of aluminum, tin foil, porcelain, or watch glass can be used for this purpose.
2. Place the representative specimen in small pieces in the container and determine the mass m_2.
3. Place the specimen and the container in an oven at 105 to 110°C and dry to constant mass m_3. The specimen is deemed to have reached a constant mass if the mass loss between two successive measurements is less than 0.1% of the initial wet mass. This can take 12 to 18 hours, and, for convenience, the samples are often left in the oven for 24 hours. A pair of tongs and/or heavy gloves can be used for removing the specimen containers from the oven (The old asbestos gloves are not welcomed in laboratories anymore!). The specimen can be allowed to cool in a desiccator for 30 minutes before determining the mass.

Notes:

1. Organic soils that can decompose and soils containing gypsum that can dehydrate should be heated at lower temperatures and for a longer period if necessary. It is common practice to maintain the oven at 60°C for organic soils.

2. If there is a delay in transferring the container with the moist sample to the oven, it can collect moisture from the air. To avoid this, use containers with lids and remove the lids (and place at the bottom of the container) when placing inside the oven.

Depending on whether the water content will be recorded to the nearest 0.1 or 1%, the minimum wet mass recommended for the water content test is given in Table B3.1.

Datasheet:

A simple datasheet for this test, prepared in Excel, is shown in Table B3.2. The water content should be reported to the nearest 0.1% for water content less than 50%, 0.5% for water content between 50 and 100%, and 1% for water content greater than 100%. The other data recorded may include:

- Sample source
- Standard followed and deviations, if any
- Any material excluded from the test
- Whether the test specimen mass was less than what is given in Table B3.1
- Drying method and temperature

Table B3.1 Minimum wet mass for water content measurements

Largest grain size, mm (sieve no.)	To the nearest 0.1%	To the nearest 1%
< 2.0 (No. 10)	20 g	20 g
4.75 (No. 4)	100 g	20 g
9.5	500 g	50 g
19.0	2.5 kg	250 g
37.5	10 kg	1 kg
75.0	50 kg	5 kg

Table B3.2 Water content measurements

Sample no.	Tin no.	m_1 (g)	m_2 (g)	m_3 (g)	w (%)	Comments
JC001	12	49.25	74.21	68.23	31.5	24 hrs in oven
JC002	18	51.04	81.27	73.56	34.2	24 hrs in oven
BP018	A3	60.34	189.2	173.6	13.8	Clayey sand; 28 hours in oven

m_1 = Mass of container

m_2 = Mass of container + specimen

m_3 = Mass of container + dry specimen

Analysis:

The *water content of the soil specimen* is defined as:

$$w(\%) = \frac{m_2 - m_3}{m_3 - m_1} \times 100 \qquad\qquad (\text{B3.1})$$

Cost: US$5–US$10

B4   Specific Gravity of Soil Grains

Objective: To determine the specific gravity (particle density) of the soil grains

Standards: ASTM D854
              AS 1289.3.5.1
              BS1377-2

Introduction

Specific gravity of a substance is simply how many times heavier it is than water. Here, water is considered the reference material that has the density of 1.00 g/cm^3 at 4°C and reduces slightly with an increase in temperature, dropping to 0.99821 g/cm^3 at 20°C, which is commonly used as the reference temperature for laboratory tests. Specific gravity, a dimensionless number, denoted by G_s, is defined as:

$$G_s = \frac{\text{Density of soil grain}}{\text{Density of water}} \qquad (B4.1)$$

Specific gravity is required in the phase relation calculations and, hence, in most laboratory tests. It lies in the range of 2.6 to 2.8 for most inorganic soils. Organic soils, fly ash, and porous particles such as diatomaceous earth may have specific gravity less than 2.0. On the other hand, mine tailing rich in minerals such as iron can have specific gravity as high as 4.0. In unit of g/cm^3, the absolute value of the *soil particle density* has the same magnitude as the specific gravity.

   Pycnometer is a glass bottle of controlled volume in the form of a stoppered volumetric flask, density bottle, or iodine flask that can be filled to a specific volume (e.g., 250 or 500 ml). It is used to determine the specific gravity of a wet or dry sample. In cohesive soils, a small sample with an equivalent dry mass of 30 to 50 g is recommended. In coarse grained soils, a larger sample with an equivalent dry mass of about 200 g and larger pycnometers are suggested. In the case of a wet sample, the water content w can be determined separately, and the dry mass m_s can be determined from the wet mass m_t as $m_t/(1 + w)$, where w is expressed in decimal and not percentage. A better way to determine the dry mass is by drying out the sample slurry at the end of the test, ensuring no grains are lost.

Procedure:

The steps involved in a specific gravity measurement are illustrated in Figure B4.1. The procedures include:

1.   Wash the density bottle with water, rinse using acetone or alcohol, and dry the bottle by blowing air. Determine the mass of the bottle with stopper, m_1. This measurement

is not required for the computations. It is used only in calibrating the pycnometer to ensure that:

$$m_3 = m_1 + \rho_w V \qquad (B4.2)$$

where V = control volume of the pycnometer (or density bottle), m_3 = mass of the pycnometer (or density bottle) filled with water, and ρ_w = density of water. Lambe (1951) and Germaine and Germaine (2009) recommend developing a calibration curve where the mass of the pycnometer m_3 with water is plotted against the temperature, and m_3 for the appropriate temperature is read from this plot.

2. Transfer the test sample into the pycnometer carefully rinsing all remaining soil particles in it using a wash bottle. Fill with more water—up to about half of the volume. Shake the bottle with stopper and agitate the water to form slurry.
3. De-air the slurry by applying vacuum, heat (boiling), or a combination of both. Agitate the bottle during this process.
4. Fill the remaining volume of the pycnometer with de-aired water and determine the mass m_2.
5. Empty the pycnometer into an evaporating dish, ensuring that all the particles are transferred. Use wash bottle with squirt to wash all the grains into the evaporating dish, leaving the pycnometer clean. A few missing grains can affect the second decimal of the specific gravity value.
6. Place the evaporating dish in an oven at 105 to 110°C for 24 hours and dry to constant mass. From this the mass of the dry soil m_s can be determined.
7. Fill the pycnometer with water and determine the mass m_3. Alternately read off m_3 from the calibration curve (see Step 1).

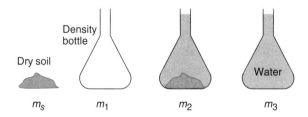

Figure B4.1 Specific gravity test

Datasheet:

A simple datasheet for this test, prepared in Excel, is shown in Table B4.1. The other data that may be included are:

- Identification of the soil (i.e., borehole, depth, and sample numbers)
- Visual description of the soil

- Percent passing 4.75 mm sieve
- Test temperatures to the nearest 0.1°C
- The average value of the readings
- Standard followed and deviations, if any
- Average specific gravity at 20°C to the nearest 0.01
- Comparison with typical values

Table B4.1 Specific gravity measurements

Sample description:	Dark brown sandy clay with high plasticity (CH)		
Sample location:	Kirwan Hospital		
Sample no.:	TP9-3		
Date:	12-March-2005		
Tested by:	Warren O'Donnell		
Notes:			
Test no.	1	2	3
Pycnometer no.	5	5	5
Mass of bottle + water, m_3 (g)	672.26	672.26	672.26
Mass of bottle + soil + water, m_2 (g)	707.88	705.98	706.32
Evaporating dish number	A	B	C
Mass of evaporating dish (g)	445.32	449.24	452.67
Mass of evaporating dish and dry soil (g)	501.34	502.35	506.21
Mass of dry soil, m_s (g)	56.02	53.11	53.54
Test temperature (°C)	25	25	25
Density of water (g/cm^3)	0.99705	0.99705	0.99705
Specific gravity @ above temperature	2.75	2.74	2.75
Specific gravity @ 20°C	2.74	2.74	2.75

Analysis:

The specific gravity of the soil grains is defined as:

$$G_s = \frac{m_s}{m_s + m_3 - m_2} \qquad (B4.3)$$

where m_s = mass of the dry soil (Step 6), m_2 = mass of pycnometer, soil and water (Step 4), and m_3 = mass of pycnometer filled with only water (Step 7). The above value of G_s should be corrected to 20°C by multiplying by $\rho_w/\rho_{w,20}$ where ρ_w and $\rho_{w,20}$ are the densities of the water at laboratory temperature and 20°C respectively. The densities of water at different temperatures are given in Table B4.2. The computed values of G_s should be within 0.03. The average of the values lying within 0.03 should be presented to two decimals as G_s. In soils containing water

Table B4.2 Density of water at different temperatures

Temperature, °C	15.0	16.0	17.0	18.0	19.0	20.0	21.0	22.0
Density, g/cm³	0.99910	0.99895	0.99878	0.99860	0.99841	0.99821	0.99799	0.99777
Temperature, °C	23.0	24.0	25.0	26.0	27.0	28.0	29.0	30.0
Density, g/cm³	0.99754	0.99730	0.99705	0.99679	0.99652	0.99624	0.99595	0.99565

soluble salts, other liquids such as kerosene may be used instead of water, and Equation B4.3 should be modified as:

$$G_s = \frac{m_s}{m_s + m_3 - m_2} \times \frac{\rho_l}{\rho_w} \qquad (B4.4)$$

where ρ_l is the density of the liquid. Some typical specific gravity values are given in Table B4.3.

Table B4.3 Typical values of specific gravity (after Day 2001, Bowles 1986, Lambe and Whitman 1979)

Soil type	G_s
Peat	1.0 or less
Organic clays	Varies; 2.0 or less
Serpentine	2.2–2.7
Attapulgite	2.30
Gypsum	2.3–2.4
Halloysite (2 H_2O)	2.55
K-feldspar	2.54–2.57
Na-Ca-feldspar	2.62–2.76
Kaolinite	2.61–2.66
Quartz sand	2.65
Silty sand	2.67–2.70
Inorganic clays	2.70–2.80
Calcite	2.70–2.72
Chlorite	2.6–2.9
Illite	2.60–2.86
Montmorillonite	2.4–2.8
Pyrophyllite	2.84
Dolomite	2.85
Soils with mica or iron	2.75–3.00
Muscovite	2.7–3.1
Biotite	2.8–3.2
Hematite	5.20–5.30
Galena	7.4–7.6

Cost: US$70–US$80

B5   Sieve Analysis

Objective: To determine the grain size distribution of the coarse fraction of a soil by sieve analysis

Standards: ASTM D6913
AS 1289.3.6.1
BS 1377-2

Introduction

The relative proportions of the different grain sizes in a soil are generally presented in the form of grain size distribution. It is generally presented graphically, where the cumulative percentage finer than a specific grain size (*y*-axis) is plotted against the grain size (*x*-axis in log scale) in mm. The coarse fraction of the soil sample is generally studied by *sieve analysis* and the fine fraction is studied by *hydrometer analysis*. In a sieve analysis, a set of sieves (Figure B5.1a) are stacked on top of each other with the opening sizes increasing upward. The soil sample is passed from the top through the stack of sieves and agitated manually or by using a mechanical sieve shaker (Figure B5.1b). From the masses of the soil fractions retained on the sieves, the grain size distribution is determined.

The *sieve* is a 200/300/450 mm diameter tray with a wire mesh at the bottom. A 25 mm sieve has 25 mm × 25 mm square openings. The opening size of the mesh varies from 75 mm

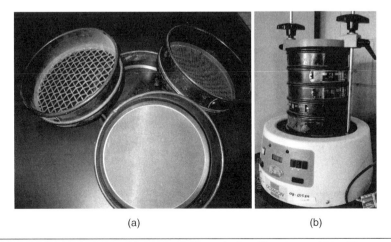

(a)   (b)

Figure B5.1 (a) Sieves and bottom pan (b) stack of sieve shaker (Courtesy of Ms. Dhanya Ganesalingam, James Cook University)

to 0.038 mm. Sieves with openings smaller than 9.5 mm (⅜ in.) have been assigned a unique number, which is simply the number of openings per inch. For example, a No. 200 sieve (0.075 mm openings) has 200 × 200 openings per square inch. It is becoming more common to refer to all sieves by the aperture size. There are slightly different standard sieves used in practice. They include ASTM (D6913), Tyler, U.S. Bureau of Reclamation, British Standard, and International Standards Organisation, among others. Some of the common sieve sizes are given in Table B5.1. Some commonly used ASTM, AS, and BS sieves are listed in Table B5.2.

Procedure:

1. Air dry the soil and determine the *hygroscopic water content* of the air-dried soil. This would generally be small—0 to 2%. Hygroscopic water content is the initial water content of the soil when it is in equilibrium with the atmosphere. This may be used to correct the retained masses to absolute dry masses.

2. A set of 200 mm diameter sieves are often adequate for the sample sizes commonly used. With larger samples (see Step 4), it may be necessary to go for the larger 300 mm or 450 mm sieves. The coarsest sieve should allow 100% of the grains to pass. Select a number of sieves (e.g., Table B5.2) such that the openings are doubled between successive sieves when going upward. This will spread the points evenly in the plot where the grain size is in log scale. Determine and record the mass of each sieve and the bottom pan to the nearest 0.1 g.

Table B5.1 Some common aperture sizes and corresponding sieve numbers

75 mm (3 in.)	2.00 mm (No. 10)	0.300 mm (No. 50)
50 mm (2 in.)	1.40 mm (No. 14)	0.250 mm (No. 60)
37.5 mm (1½ in.)	1.18 mm (No. 16)	0.150 mm (No. 100)
25 mm (1 in.)	1.00 mm (No. 18)	0.106 mm (No. 140)
19 mm (¾ in.)	0.85 mm (No. 20)	0.090 mm (No. 170)
9.5 mm (⅜ in.)	0.60 mm (No. 30)	0.075 mm (No. 200)
4.75 mm (No. 4)	0.50 mm (No. 35)	0.063 mm (No. 230)
2.36 mm (No. 8)	0.425 mm (No. 40)	0.038 mm (No. 400)

Table B5.2 ASTM, AS, and BS sieve sizes (mm)

ASTM	AS	BS
76 (3″); 51 (2″); 38 (1½″); 25 (1″); 19 (¾″); 9.5 (⅜″); 4.75 (#4); 2.0 (#10); 0.850 (#20); 0.425 (#40); 0.250 (#60); 0.150 (#100); 0.106 (#140); 0.075 (#200).	75; 63.0; 37.5; 26.5; 19; 13.2; 9.50; 6.70; 4.75; 2.36; 1.18; 0.600; 0.425; 0.300; 0.212; 0.150; 0.075.	75; 63; 50; 37.5; 28; 20; 14; 10; 6.3; 5; 3.35; 2; 1.18; 0.600; 0.425; 0.300; 0.212; 0.150; 0.063.

3. Make a nest of sieves by stacking them on top of each other such that the aperture sizes increase from bottom to top. The pan is at the bottom of the stack to collect the material passing the finest sieve. For greater precision, the sieving process can be carried out in three stages—coarse, intermediate, and fines separated at 19.0 mm and 2.36 mm sieves, respectively.

4. The million dollar question here is how much soil to sieve. Germaine and Germaine (2009) suggest a simple rule. For a reporting resolution of 1%, the dry mass of the sample should be more than 100 times the mass of the largest grain. For 0.1% resolution, it has to be 1000 times. For example, assuming the largest grain to be 25 mm spheres with $G_s = 2.65$ and mass of 21.7 g, we should sieve 2170 g for 1% resolution and 21.70 kg for 0.1% resolution. ASTM and BS recommend a minimum mass of sample for sieve analysis.

5. Break the soil clusters into individual grains with fingers or a rubber-tipped pestle.

6. Guidelines are often given to avoid overloading the sieves. Those specified by AS 1289.3.6.1 are given in Table B5.3 which suggests the maximum mass that can be allowed on a sieve. Similar values are also given in ASTM D6913 and BS 1377-2.

Table B5.3 Maximum mass of material to be retained on each sieve (AS 1289.3.6.1)

Aperture size (mm)	Maximum mass of soil to be retained (g)		
	200 mm dia.	300 mm dia.	450 mm dia.
200	—	3750	7500
75	—	3000	6000
63	—	2750	5500
37.5	1000	2200	5000
26.5	800	1800	4000
19	600	1200	3000
13.2	400	900	2000
9.5	250	600	1500
6.7	230	500	1200
4.75	200	400	1000
2.36	150	300	600
1.18	100	200	500
0.600	75	150	300
0.425	60	120	250
0.300	50	100	200
0.212	45	90	180
0.150	40	80	160
0.075	25	50	100

Note: ASTM and BS specify different values

7. Place the sample at the top of the nest of sieves and close the lid. Agitate them by hand using horizontal rotations and vertical motion, or by a mechanical sieve shaker, for 10 to 15 minutes until no more than 1% of the mass retained in any sieve will pass through the openings during the next one minute of hand sieving.

8. Determine the mass of the soil retained on each sieve. It is better to empty each sieve into a container for weighing. This will not require that the masses of the empty sieves be determined beforehand, as in the datasheet (Table B5.4). Sieve brushes can be used to remove the grains that are stuck between the wires. Wire brushes for coarse sieves and nonmetallic bristle brushes for fine sieves are recommended.

Table B5.4 Sieve analysis data

Soil description: Sand with fines
Sample location: Annandale Gardens
Sample no.:   B1
Date: 23/3/2002
Tested by: Rudd Rankine
Notes:
Mass of air-dried sample (g):   424.2
Hygroscopic water content (%):

Sieve opening (mm)	Sieve only (g)	Sieve and soil (g)	Soil retained (g)	% retained	Cumulative % retained	Cumulative % passing
9.5	By inspection—none			0	0.0	100.0
4.75	532.6	548.9	16.3	3.9	3.9	96.1
2.36	452.9	575.3	122.4	28.9	32.8	67.2
1.18	421.3	537.1	115.8	27.4	60.2	39.8
0.6	376.4	434.5	58.1	13.7	73.9	26.1
0.3	389.5	423.1	33.6	7.9	81.8	18.2
0.15	412.6	423.4	10.8	2.6	84.4	15.6
0.075	304.8	312.3	7.5	1.8	86.1	13.9
Pan	325.6	384.2	58.6	13.9	100.0	NA
	Total		423.1	100.0		

Datasheet:

A simple datasheet for this test, prepared in Excel, is shown in Table B5.4. While the principles remain the same, some simple manipulations are required to account for:

- Hygroscopic water content
- Wet sieving
- Different sieve sets (e.g., 200 mm and 300 mm) used to avoid overloading
- Multistage sieving (i.e., separate sieving of coarse, intermediate, and fine grains) when large sample sizes are involved

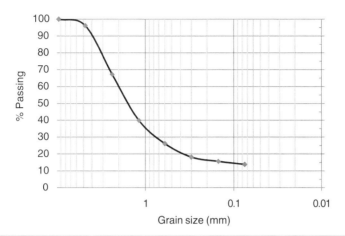

Figure B5.2 Grain size distribution curve of the coarse fraction

Analysis:

Computation of the cumulative percentages passing different sieve sizes is fairly straightforward. The results are presented in the form of a grain size distribution curve as shown in Figure B5.2. The grain size distribution of the fine fraction passing 0.075 mm (No. 200) sieve is studied using hydrometer analysis (see Test B6). The data from hydrometer analysis can be combined with the sieve analysis data in completing the grain size distribution plot (see Figure B6.3).

Cost: US$80–US$120

B6   Hydrometer Analysis

Objective: To determine the grain size distribution of the fine fraction of a soil by hydrometer analysis

Standards: ASTM D422
AS 1289.3.6.3
BS 1377-2

Introduction

The relative proportions of the different grain sizes present in a fine grained soil are determined through a hydrometer test. The test is based on Stokes' law where the terminal velocity v of a spherical grain falling through water can be related to the grain size D by:

$$D = \sqrt{\frac{18\,\mu_w\,v}{\gamma_s - \gamma_w}} \tag{B6.1}$$

where μ_w = dynamic viscosity of water, v = terminal velocity of the spherical grain, γ_s = unit weight of the grain (= $G_s\gamma_w$), and G_s = specific gravity of the soil grains. This equation is valid for D in the range of 0.0002 mm to 0.2 mm.

A hydrometer (Figure B6.1) is used to measure the density of a liquid in a wide range of applications and works on Archimedes' principle. While floating in the liquid, it measures the density near the center of the bulb that is at a depth H_R (known as effective depth) below the water level. When the hydrometer is placed in a soil-water suspension at time t, any soil grain larger than a certain diameter D would have fallen below the center of the bulb. Since H_R is the maximum distance through which these grains have fallen during time t, the velocity can be written as:

$$v = \frac{H_R}{t} \tag{B6.2}$$

Substituting Equation B6.2 in Equation B6.1:

$$D(\text{mm}) = \sqrt{\frac{30\,\mu_w\,(\text{Poise})}{980\,(\rho_s - \rho_w)}} \sqrt{\frac{H_R(\text{cm})}{t(\text{min})}} = K\sqrt{\frac{H_R(\text{cm})}{t(\text{min})}} \tag{B6.3}$$

where ρ_s and ρ_w are the soil particle density and the density of water, respectively, in g/cm^3. For water at 20°C, $\mu_w \approx 0.01$ poise. From Equation B6.3, the grain size corresponding to time t can be determined. The values of K at various temperatures, based on the appropriate values of μ_w and ρ_w, are given in Figure B6.2.

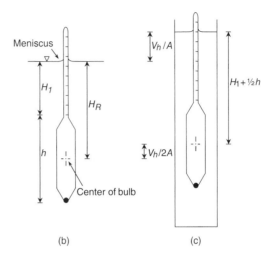

(a)

Figure B6.1 Hydrometer: (a) photograph of a hydrometer in water; (b) in suspension permanently; and (c) inserted into suspension only during reading (Photograph: N. Sivakugan)

The hydrometer measures the density of the suspension at the center of the bulb that is a measure of the mass of soil grains present in the suspension (Figure B6.1). At depth H_R, the suspension is free of any grains larger than D. The density reading can be translated into the corresponding percentage passing.

Types of hydrometers:

There are two major types of ASTM recommended hydrometers—151H and 152H—that are used with soil suspensions. 151H is graduated in specific gravity values from 0.995 to 1.038 with divisions of 0.001 (i.e., four significant figures). 152H is graduated in gram of soil per liter of suspension from -5 to 60 with divisions of 1.0 g/L.

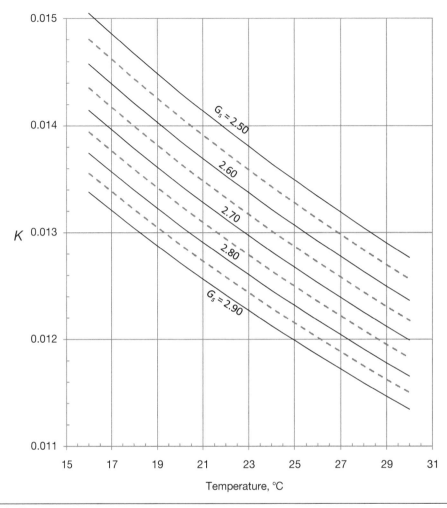

Figure B6.2 K vs. temperature

If the soil mass in suspension at time t is m (grams) and the volume of the suspension is V (ml), the density (g/mL) of the suspension is:

$$\rho_{susp}\,(g/ml) = \frac{m + \left(V - \dfrac{m}{G_S\rho_w}\right)\rho_w}{V} = \frac{m}{V} + 1 - \frac{m}{G_s V}$$

$$m = \frac{G_s}{G_s - 1} V(\rho_{susp} - 1) \tag{B6.4}$$

For $G_s = 2.65$ and $V = 1000$ mL, $\rho_{susp} = 1.000$ g/mL (in 151H) corresponds to $m = 0$ g (in 152H) and $\rho_{susp} = 1.0374$ g/mL (in 151H) corresponds to $m = 60$ g (in 152H). Equation B6.4 relates the readings between the two types of hydrometers. The analysis below is based on a 152H-type hydrometer.

The hydrometer can remain in the suspension throughout the test or it can be placed in the suspension while taking the reading every time. When it is placed in the suspension permanently, the effective depth H_R is given by (Figure B6.1b):

$$H_R = H_1 + \frac{h}{2} \qquad (B6.5)$$

where h = length of the hydrometer bulb (14 cm for 152H type) and H_1 = depth from the surface of the suspension to the top of the bulb. When the hydrometer is inserted into the suspension for every reading (Figure B6.1c), the surface of the suspension rises by V_h/A where V_h = volume of the hydrometer (ml) and A = cross section of the volumetric cylinder (cm^2). The center of the bulb is thus displaced upward by $V_h/2A$, and the effective depth H_R is given by:

$$H_R = H_1 + \frac{h}{2} - \frac{V_h}{2A} = H_1 + \frac{1}{2}\left(h - \frac{V_h}{A}\right) \qquad (B6.6)$$

For a standard 152H hydrometer, V_h = 67 cm^3 and A = 27.8 cm^2 for a 1000 mL graduated cylinder (Bowles 1986). The H_R values can be measured in the laboratory for several values of hydrometer reading R_h and the calibration curve can be developed. For all practical purposes this is approximately linear. For 152H-type hydrometers with the standard dimensions (h = 14 cm, V_h = 67 cm^3 and A = 27.8 cm^2; H_1 = 10.5 cm at 0 g/l, and H_1 = 2.3 cm at 50 g/L), it can be seen from Equation B6.6 that:

when R_h = 0 (g/ml) H_R = 17.5 if hydrometer is placed in the suspension permanently

H_R = 16.3 if hydrometer is placed in the suspension only for each reading

when R_h = 50 (g/ml) H_R = 9.3 if hydrometer is placed in the suspension permanently

H_R = 8.1 if hydrometer is placed in the suspension only for each reading

For situations where the hydrometer is placed in the suspension only during reading, the effective depth H_R and the hydrometer reading R_h can be related by:

$$H_R(\text{cm}) = 16.3 - 0.1641\, R_h \qquad (B6.7)$$

For the situation where the hydrometer is placed in the suspension permanently, Equation B6.7 becomes:

$$H_R(\text{cm}) = 17.5 - 0.1641\, R_h \qquad (B6.8)$$

Due to the presence of meniscus, and especially in turbid suspensions, we read R_h' at the upper rim of the meniscus and not R_h. The R_h values to be used in Equations B6.7 and B6.8, corrected for meniscus, are given by:

$$R_h = R_h' + C_m \qquad (B6.9)$$

where C_m is the meniscus correction.

Percent finer:

Two other corrections that are applied to the hydrometer reading R'_h are to account for temperature and the dispersing agent. The temperature correction C_t, for hydrometers calibrated at 20°C, is given in Table B6.1. The dispersing agent correction C_a is the reading of the hydrometer placed in a sedimentation cylinder containing a solution with the same amount of dispersing agent as in the suspension and at the same temperature. This correction, also known as zero correction, accounts for the water impurities. Remember, the hydrometer may not be reading zero when placed in water. The hydrometer reading R_c corrected for temperature and dispersing agent and meniscus is given by:

$$R_c = R'_h + C_m + C_t - C_a = R_h + C_t - C_a \tag{B6.10}$$

The value of R_c in Equation B6.10 gives the mass of grains smaller than diameter D in suspension near the center of the bulb. Therefore, if the mass of soil grains used in the suspension at the start is m_s, the corresponding percentage finer than D is given by $(R_c/m_s) \times 100\%$.

The hydrometers are calibrated for soil suspensions containing soil grains where $G_s = 2.65$. From Equation B6.4, it can be deduced that for other values of G_s the percentage finer than P is given by:

$$P = \frac{R_c \times a}{m_s} \times 100\% \tag{B6.11}$$

where:

$$a = \left(\frac{G_s}{G_s - 1}\right)\left(\frac{2.65 - 1}{2.65}\right) = 0.6226\left(\frac{G_s}{G_s - 1}\right) \tag{B6.12}$$

For 151H-type hydrometers:

$$P = \frac{100,000}{m_s}\left(\frac{G_s}{G_s - 1}\right)(R - 1) \tag{B6.13}$$

where R is the corrected hydrometer reading.

It must be remembered that the sample used in the hydrometer test is a small fraction of the soil passing the 75 μm (No. 200) sieve and retained on the pan. Therefore, the percentage finer than P determined from Equation B6.11 or B6.13 should be translated to the whole soil sample that was being sieved. If m_{fines} = mass of the fines retained on the pan, and m_{whole} = mass of the

Table B6.1 Temperature correction C_t for hydrometers calibrated at 20°C

Temp °C	15	16	17	18	19	20	21	22
C_t	−1.10	−0.90	−0.70	−0.50	−0.30	0.0	+0.20	+0.40
Temp °C	23	24	25	26	27	28	29	30
C_t	+0.70	+1.00	+1.30	+1.65	+2.00	+2.50	+3.05	+3.80

whole soil sample used in the sieve analysis, the above values in Equations B6.11 and B6.13 should be multiplied by m_{fines}/m_{whole}.

Procedure:

1. *Dispersing agent*: A dispersing agent is required to ensure that the grains in the suspension do not flocculate, thus forming a large floc that settles faster. Sodium hexametaphosphate ($NaPO_3$—trade name Calgon) is the commonly used dispersing agent. ASTM D422 suggests mixing 40 g of sodium hexametaphosphate with distilled or demineralized water to make 1000 mL solution. The stock solution should be stored in a dark bottle and out of sunlight for later use (up to a month).

2. *Dispersion of soil sample*: Determine the mass of about 50 g of air-dried soil and mix it with 125 mL of dispersing agent solution. Stir this until all the soil grains are thoroughly wetted and soak for at least 16 hours. At the end of soaking period, disperse the sample further by adding more dispersing agent solution and stirring in the special dispersion cup. Transfer the content to a 1000 mL glass sedimentation cylinder and add distilled or demineralized water until the total volume of the suspension is 1000 mL. Seal the open end of the cylinder with a rubber stopper and turn the cylinder upside down and back for one minute to ensure that all the grains are in the suspension. Place the cylinder in a constant-temperature bath.

3. Take hydrometer readings at 0.5, 1, 2, 4, 8, 15, and 30 minutes, 1, 2, 4, 8, 24 and 48 hours. For the early readings, it may be easier to leave the hydrometer within the suspension (use Equation B6.8 for H_R) and later it can be placed in the suspension a few seconds prior to reading (use Equation B6.7 for H_R). Note that R_h is the hydrometer reading and H_R is the depth of the center of the bulb in the suspension.

4. *Meniscus correction, C_m*: Due to turbidity of the suspension, read the top rim of the meniscus R'_h and add the meniscus correction C_m for the correct level of the surface of the suspension. The meniscus correction is simply the difference between the top and bottom of the meniscus when the hydrometer is placed in water.

5. *Temperature correction, C_t*: The hydrometer was calibrated at 20°C by the manufacturer. Noting that the density and viscosity of water depends on temperature, the hydrometer readings would be affected by the temperature. Add the temperature correction C_t given in Table B6.1 to the hydrometer reading.

6. *Zero correction or dispersing agent correction, C_a*: Mix the same amount of dispersing agent with the same source of water, making 1000 mL of solution. Place the hydrometer in the solution and note the reading as the zero or dispersing agent correction C_a. This reading should be subtracted from the hydrometer reading.

7. *K-factor*: Determine the factor K from Figure B6.2.

Notes:

1. For type 151H that reads the density of the suspension in g/mL, the dispersing agent correction C_a is obtained by subtracting 1.0 from the hydrometer reading.
2. ASTM D422 and most laboratory manuals recommend placing the hydrometer in the suspension only while taking the reading, not permanently. However, for the first few readings that are taken at short time intervals, it is convenient to leave the hydrometer in the suspension, which is allowed in the Australian Standards.

Table B6.2 Hydrometer data

Soil description:	Dredged mud		K	0.0138		$m_s(g)$	46.3
Sample location:	Port of Brisbane		C_m	1		$m_{fines}(g)$	58.6
Date:	14th May 2010		C_t	0.4		$m_{whole}(g)$	424.2
Test procedure:	ASTM D422		C_a	3			
Hydrometer No:	152H–B4969		*Comments:*				
Sp. Gravity G_s	2.65	*a* 0.9999					
Dispersing agent (amount and type): 125 mL of sodium hexa-metaphosphate (40 g/1000 mL)							

Date	Time of Reading	Elapsed time (min)	Temp (°C)	Hydrometer reading R'_h	$R_h = R'_h + C_m$	$R_c = R_h + C_t - C_a$	H_R (cm) from Eqs. B6.7/6.8	D (mm)	% finer Hyd. Sample	% finer Whole sample
14-May-2010	8:55:00	0.5	22	46.5	47.5	44.9	8.5	0.05692	97.0	13.4
	8:56:00	1	22	46	47	44.4	8.6	0.04044	95.9	13.2
	8:57:00	2	22	44.5	45.5	42.9	8.8	0.02900	92.7	12.8
	8:59:00	4	22	43	44	41.4	9.1	0.02079	89.4	12.4
	9:03:00	8	22	40	41	38.4	9.6	0.01510	82.9	11.5
	9:10:00	15	22	37.5	38.5	35.9	10.0	0.01126	77.5	10.7
	9:25:00	30	22	34.5	35.5	32.9	10.5	0.00815	71.1	9.8
	9:55:00	60	22	31.5	32.5	29.9	11.0	0.00590	64.6	8.9
	10:55:00	120	22	28	29	26.4	11.5	0.00428	57.0	7.9
	12:55:00	240	22	25	26	23.4	12.0	0.00309	50.5	7.0
	16:55:00	480	22	22.5	23.5	20.9	12.4	0.00222	45.1	6.2
15-May-2010	8:55:00	1440	22	19	20	17.4	13.0	0.00131	37.6	5.2
	16:55:00	1920	22	18	19	16.4	13.2	0.00114	35.4	4.9
16-May-2010	8:55:00	2880	22	17	18	15.4	13.3	0.00094	33.3	4.6

Datasheet:

A simple datasheet for a hydrometer test, using type 152H hydrometer is shown in Table B6.2. From the sieve analysis (see Test B5), there were 58.6 g of fines in the whole sample of 424.2 g. Out of this 58.6 g, only a 46.3 g sample was used in the hydrometer test.

Analysis:

The theoretical developments in a hydrometer test are discussed in the introduction. The computations in a hydrometer test are fairly straightforward. The hydrometer depth H_R is determined from the hydrometer reading that is corrected only for meniscus. This can be done using the calibration between H_R and R_h developed in the laboratory or Equations B6.7 and B6.8 that are valid for the standard 152H hydrometer. At a specific time, the grain size D is determined from Equation B6.3. Corresponding percentage finer than this diameter is obtained from Equation B6.11 where R_c is the hydrometer reading corrected for meniscus, temperature and dispersing agent.

The grain size distribution data for the hydrometer sample has to be translated to the larger sample from which the fines were obtained for the hydrometer test (see Figure B6.3).

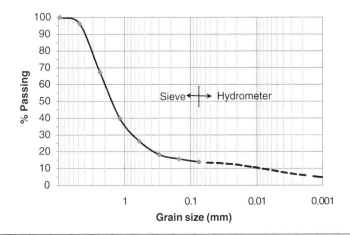

Figure B6.3 Grain size distribution curve

Cost: US$70–US$100

B7 pH

Objective: Determination of pH value by the electrometric method

Standards: ASTM D4972
AS 1289.4.3.1*
BS 1377-3*

Introduction

Water (H_2O) consists of hydrogen ions (H^+) and hydroxyl ions (OH^-) that can be present as free ions carrying electric charges. Distilled water has H^+ concentration of 10^{-7} g/L. Addition of acid to distilled water increases the concentration of H^+ and decreases the concentration of OH^-. Since the concentrations are very small, they are expressed in a logarithmic scale. The pH value is the negative of the logarithm (base 10) of the concentration of hydrogen ions in an aqueous solution. The "p" and "H" stand for the mathematical power and hydrogen ions respectively. It is a dimensionless number that provides a measure of the acidity (pH of 0 to 7) or alkalinity (pH of 7 to 14) of the solution on a scale reading from 0 to 14, where 7 represents neutrality. A reduction of 1 unit of pH implies an increase of the hydrogen ion concentration in the solution by 10 times. ASTM D4972 suggests two methods for measuring pH: Method A using pH sensitive electrode (Figure B7.1a) and Method B using pH sensitive paper. Method A is more accurate, which can measure pH to 0.05 and is discussed here. The tests are performed within the temperature range of 15 to 25°C.

Buffer solutions are used for calibration of the pH meter. A buffer solution with a pH within 2 units of soil being tested should be used. Two suggested buffer solutions are:

1. pH = 4.0: Dissolve 5.106 g of potassium hydrogen phthalate in distilled water to make 500 mL.
2. pH = 9.2: Dissolve 9.54 g of sodium tetraborate (borax) in distilled water to make 500 mL.

In addition, a saturated solution of potassium chloride is required for holding the calomel electrode (reference electrode).

Typically, soils have a pH in the range of 3.5 to 9.0.

Procedure:

1. From the soil passing a 2.36 mm sieve (2.00 mm in ASTM and BS), obtain a sample of about 35 g.
2. Transfer exactly 30 g of the above sample into a 100 mL beaker and add 75 mL of deionized or distilled water. Stir the suspension for a few minutes, then cover the beaker with

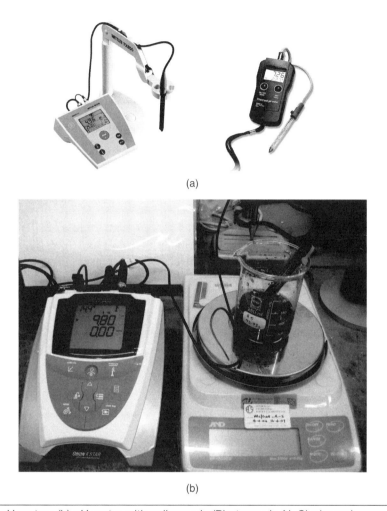

(a)

(b)

Figure B7.1 (a) pH meters (b) pH meter with soil sample (Photograph: N. Sivakugan)

a cover glass and allow it to stand for several hours. Stir it again immediately before testing. For stabilized soils, the standing time should not be more than 2 h, otherwise irreversible changes may take place in the stabilizers.

3. Calibrate the pH meter by placing the electrodes in the standard buffer solutions (e.g., pH of 4.0 and 9.2) separately, following the procedure recommended by the manufacturer.

4. Wash the electrodes with distilled water and immerse in the soil-suspension (Figure B7.1b). Take two or three readings of the pH of the soil-suspension with brief stirring between each reading. These readings should not differ by more than 0.05 pH units.

5. Remove the electrodes from the suspension and wash with deionized or distilled water. Check the calibration of the pH meter against one of the standard buffer solutions. If

the instrument is out of adjustment by more than 0.05 pH units, set it to the correct adjustment and repeat the procedure given in Step 4 until consistent readings within 0.1 units are obtained.

6. When not in use, leave the electrodes standing in a beaker of deionized or distilled water.

Note: ASTM D4972 suggests measuring pH separately in soil suspensions made using a water and calcium chloride solution and reporting both values. The pH in calcium chloride suspension is slightly less.

Datasheet:

A simple datasheet for this test, prepared in Excel, is shown in Table B7.1. The pH value should be reported to the first decimal place. The other data recorded may include:

- Sample source
- Sampling date
- pH meter used
- Temperature of the suspension while reading pH
- Standard followed and deviations, if any

Table B7.1 pH datasheet

Soil description: Low plastic silty clay
Sample location: Clayton campus
Date: 21 May 2010   Test procedure: AS 1289.4.3.1
Notes:

Sample no.	Test 1	Test 2	Test 3	Test 4	Test 5	Average pH	Temp°C
BH2-2.2m	9.80	9.81	9.79	9.78	9.82	9.80	23
TP2-1.5m	7.30	7.33	7.32	7.31	7.32	7.32	23
TP8-2.0m	8.45	8.44	8.45	8.44	8.44	8.44	23
BH7-1.5m	8.20	8.21	8.22	8.21	8.21	8.21	23

Analysis:

There is no analysis required for this test.

Cost: US$20–US$25

B8   Organic Content

Objective: To determine the organic content of soils and aggregates

Standards: ASTM D2974
BS 1377-3

Introduction

Soils and aggregates can contain organic matter to varying levels. The organic content of aggregates and other organic soils such as peats, organic clays, silts, and mucks can be determined by loss on ignition method. The oven-dried soil is baked in a muffle furnace at very high temperatures to burn off the organic matter, leaving the inorganic ash (soil fraction that is not burnt) behind. *Ash content* is the percentage of the dry mass that is left behind, and *organic content* is the percentage that is lost on ignition. They add up to 100%. The organic content test is generally carried out on relatively small specimens with dry mass of as low as 10 g. In road aggregates and structural fills that have low organics content (e.g., 0 to 5%), larger specimens are preferable. An alternate method of determining organic content is by chemical oxidation.

Procedure:

1. Determine the mass of a covered high silica or porcelain dish to the nearest 0.01 g.
2. Place the oven-dried (at 105 to 110°C) test specimen in the dish and determine the mass of the specimen m_1 to the nearest 0.01 g (Figure B8.1).

Figure B8.1 Porcelain dish with recycled glass samples (Photograph courtesy of Mahdi Disfani, Swinburne University of Technology, Australia)

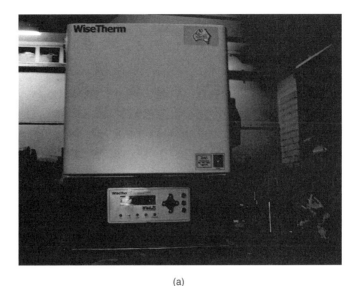

(a)

(b)

Figure B8.2 (a) Muffle furnace (b) with samples inside (Courtesy of Mahdi Disfani, Swinburne University of Technology, Australia)

3. Remove the cover and place the dish in a muffle furnace (Figures B8.2a and b). Gradually bring the temperature in the muffle furnace to 440°C ± 22°C and hold for 4 to 5 hours until the specimen is completely ashed with no change of mass after at least 1 hour of heating.

4. Remove the dish from the furnace and cover with aluminum and cool in a desiccator.

5. Determine the mass m_2 of the ash to the nearest 0.01 g.

Datasheet:

The organic content and ash content should be reported to the nearest 0.1%. A simple datasheet for this test, prepared in Excel, is shown in Table B8.1. The other data recorded may include the following:

- Sample source
- Sampling date
- Muffle furnace temperature used in ash content determination
- Standard followed and deviations, if any

Table B8.1 Organic content datasheet

Soil description: Clayey, sandy gravel
Sample location: Clayton campus
Date: 21 May 2010   Test procedure: ASTM D2974-07a
Notes: Four samples from bag A

Sample no.	A1	A2	A3	A4
Dish number	1	2	3	4
Mass of dish (g)	82.95	80.80	84.45	80.80
Mass of dish and oven-dried specimen (g)	284.29	281.57	284.89	280.82
Mass of dish and ash (g)	283.30	280.65	283.95	279.78
Mass of oven-dried specimen, m_1(g)	201.34	200.77	200.44	200.02
Mass of ash, m_2(g)	200.35	199.85	199.50	198.98
Ash Content (%)	99.51	99.54	99.53	99.48
Organic content (%)	0.49	0.46	0.47	0.52

Analysis:

Ash content is calculated to the nearest 0.1% as:

$$\text{Ash content (\%)} = \frac{m_2}{m_1} \times 100 \qquad (B8.1)$$

where:

$$m_1 = \text{mass (g) of oven-dried specimen}$$
$$m_2 = \text{mass (g) of ash}$$

The organic content is determined to the nearest 0.1 % as:

$$\text{Organic content (\%)} = 100 - \text{Ash content (\%)} \qquad (B8.2)$$

Cost: US$50–US$60

B9   Liquid Limit—Casagrande's Percussion Cup Method

Objective: To determine the liquid limit of a fine grained soil using Casagrande's percussion cup method

Standards: ASTM D4318
AS 1289.3.1.1
BS 1377- 2

Introduction

The Atterberg limits were introduced in 1911 by a Swedish scientist working with ceramics and pottery. He noted that the consistency of clays depends on the water content, and he defined several borderline water contents where some noticeable change in consistency takes place. Out of the six limits he defined, only three are relevant in geotechnical context, which were later modified by Casagrande (1932). They are: *shrinkage limit*, *plastic limit*, and *liquid limit*. They are used in classifying fine grained soils. In addition, they have been related to strength, compressibility, and other geotechnical properties of clays.

When the water content of a clay is increased, the shear strength decreases. Beyond certain water content, the clay will simply flow like a liquid. That water content is known as the liquid limit denoted by LL or w_L. At the liquid limit, the shear strength is approximately 2 kPa. The liquid limit can be determined by one of the following two standard procedures:

1. Casagrande's percussion cup method
2. Fall cone method

Casagrande's apparatus is shown in Figure B9.1a. The dimensions of the apparatus and complete details are given in ASTM D4318 and other standards. Here, a moist pat of remolded fine grained soil is placed in a standard metal cup, and the surface is made flat using a spatula or palette knife. A straight V-groove is cut into the soil pat, separating it into two halves (Figure B9.1b). The cup is raised and dropped by 10 mm onto the base (Figure B9.1a), and the number of blows required to close the groove by 13 mm (½ in.) is recorded (see Figure B9.1c). The test is repeated at four or more water contents. The water content corresponding to 25 blows is defined as the liquid limit.

Procedure:

1. Obtain 250 g of soil passing a No. 40 (425 μm) sieve.
2. Place the entire sample in a mixing bowl and add water gradually to make a thick homogeneous paste by mixing thoroughly. Leave this mixture to cure and for the moisture to equilibrate for 12 to 24 hours.

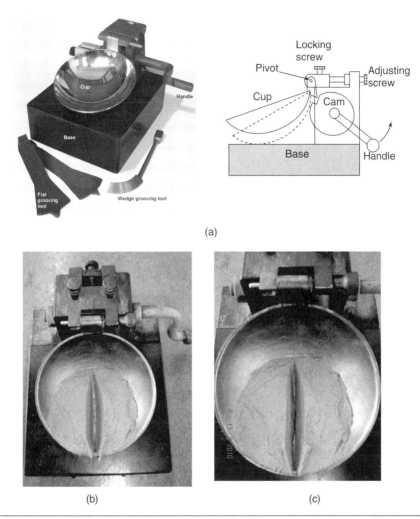

(a)

(b)   (c)

Figure B9.1 (a) Casagrande's percussion cup apparatus with grooving tools, (b) soil paste with groove, and (c) groove closure over 12 mm length (Photograph: N. Sivakugan)

3. Place the entire sample in a mixing bowl and mix thoroughly.
4. Using a spatula or palette knife, place the mixture in the cup while it is in contact with the base and level the surface parallel to the base with a maximum thickness of 10 mm.
5. Cut a V-groove using the grooving tool that divides the soil paste into two halves.
6. Turn the crank to lift the cup by 10 mm and drop it twice every second, each drop taken as a *blow*. The groove starts closing—count the number of blows required to close the grove over 13 mm (½ in.). There is a motorized apparatus available with a counter.
7. Remove a small sample of 10 to 15 g for water content determination and return the remaining soil to the mixing bowl and remix thoroughly.

8. The water content of the mixture in the bowl can be changed by air drying or adding distilled water.
9. Repeat Steps 3 to 6 at 4 different water contents, ideally with the number of blows evenly spread in the range of 15 to 40.
10. Plot water content as the ordinate (arithmetic scale) against the number of blows as abscissa (semi-log) and draw the line of best fit. Determine the water content corresponding to 25 blows as the liquid limit.

Notes:

1. AS 1289.3.1.1 requires the groove to close over a length of 10 mm.
2. There are two different grooving tools: (1) flat grooving tool and (2) wedge grooving tool. To limit the variables, the flat grooving tool is currently recommended. The wedge grooving tool works better in sandy samples where the flat one will tear the soil pat.
3. Soil fraction passing a No. 200 (75 μm) sieve may seem more logical for the liquid limit test—remember, we are supposed to test the fines! No. 40 (425 μm), however, is the commonly accepted maximum grain size that can be used for Atterberg limits tests, where the sample contains some fine sands too.
4. The 10 mm drop should be checked prior to the test using the special height gage.
5. There are two types of base (Figure B9.1a): (1) Hard rubber, and (2) Micarta No. 221™. Most standards recommend a rubber base of a specific hardness.

Datasheet:

A simple datasheet for this test, prepared in Excel, is shown in Table B9.1. The liquid limit should be reported as an integer. The other data recorded may include:

- Sample source
- Visual classification
- Percentage of material retained on a 425 μm (No. 40) sieve
- Standard followed and deviations, if any
- History of the sample (e.g., natural state, air-dried, or oven-dried)

If it was not possible to determine either liquid limit, report the soil as non-plastic.

Table B9.1 Datasheet for Casagrande's cup method to determine liquid limit

Sample description:	Dark brown sandy clay with high plasticity (CH)
Sample location:	Kirwan Hospital
Sample No.:	TP9-3
Date:	12-March-2005
Tested by:	Warren O'Donnell
Notes:	

Test no.	1	2	3	4	5
Container no.	A3	A4	A23	A41	
Mass of container (g)	9.12	8.23	8.65	9.34	
Mass of cont. + moist soil (g)	24.31	20.84	24.12	22.53	
Mass of cont. + dry soil (g)	19.03	16.34	18.23	17.34	
Mass of water (g)	5.28	4.5	5.89	5.19	
Mass of dry soil (g)	9.91	8.11	9.58	8	
Water content (%)	53.28	55.49	61.48	64.88	
No. of blows	36	28	21	17	

Analysis:

Figure B9.2 shows the plot of water content against the number of blows. The line of best fit shown in the figure is known as the *flow curve*. The water content corresponding to $N = 25$ is read off the plot as 58, which is the liquid limit. For the diehards who wish to plot these points manually, a blank semi-log graph sheet is provided in Figure B9.3.

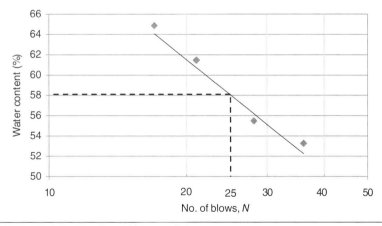

Figure B9.2 Water content vs. number of blows plot

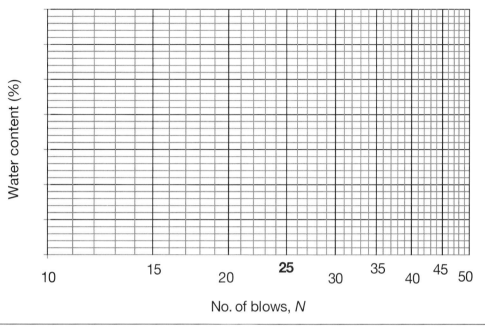

Figure B9.3 A blank semi-log graph sheet

Cost: US$70–US$80

B10   Liquid Limit—Fall Cone Method

Objective: To determine the liquid limit of a fine grained soil using the fall cone method

Standards: AS 1289.3.9.1
BS 1377-2

Introduction

Determination of the liquid limit using Casagrande's percussion cup was discussed previously. The second method for determining liquid limit is by using the fall cone that was developed by Hansbo (1957). The British Standard recommends the use of fall cone over the percussion cup for better reproducibility as it is less operator-sensitive. In addition, Casagrande's cup method introduces a dynamic effect whereas the cone method is a static test of the soil's shear strength.

The fall cone liquid limit apparatus is shown in Figure B10.1a. Here, a standard cone with a mass of 80 g (including the stem and lead shots), apex angle of 30°, and a length of 35 mm is allowed to fall (from the initial rest position when the tip is just touching the soil surface) through a soil paste contained in a standard cup of 55 mm diameter and 40 mm height over a period of 5 seconds. The penetration depth of the cone is then measured. The liquid limit is defined as the water content of the soil paste that gives 20 mm penetration.

Procedure:

1. Obtain 250 g of soil passing a 425 μm (No. 40) sieve.
2. Place the entire sample in a mixing bowl and add water gradually to make a thick homogeneous paste by mixing thoroughly. Leave this mixture to cure and for the moisture to equilibrate for 12 to 24 hours.
3. Place the entire sample in a mixing bowl and mix thoroughly.
4. Using a spatula or palette knife, fill the cup with the soil paste. Strike off the excess, leaving a smooth flat surface level with the rim of the cup.
5. Place the cup centrally under the cone. Lower the cone until it just touches the soil surface and clamp it in that position (Figure B10.1b). Set the penetration measurement device (more commonly a large dial gage) to zero.
6. Release the cone for 5 seconds, thus allowing the cone to penetrate the paste (Figure B10.1c) and re-clamp the cone.
7. Use the penetration measurement device to measure the distance the cone has penetrated into the soil paste to the nearest 0.1 mm.
8. Determine the water content.

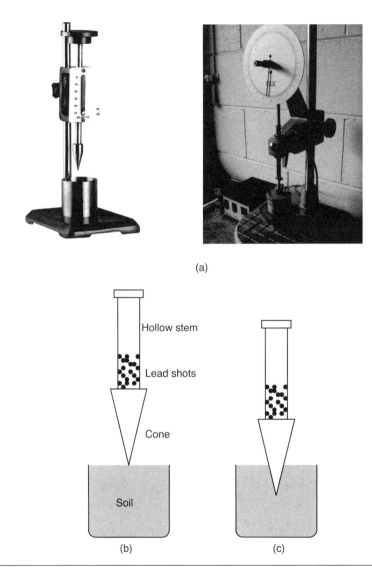

(a)

Figure B10.1 Fall cone apparatus: (a) complete apparatus; (b) cone at rest position; and (c) cone after the 5 s free fall (Photograph courtesy of Warren O'Donnell, James Cook University, Australia)

9. Repeat Steps 3 to 8 for four different water contents, ideally with two having less than 20 mm penetration and two having more than 20 mm penetration; preferably in the range of 15 to 25 mm.

10. Plot water content as the ordinate (arithmetic scale) against the penetration depth as abscissa (semi-log) and draw the line of best fit. Determine the water content corresponding to 20 mm penetration as the liquid limit.

Notes:

1. Standards recommend two determinations of penetrations at each water content and using the average value.
2. Head (1992a) compared the liquid limit values determined by the two methods and showed that there is insignificant difference between them when the liquid limit is less than 100. Above 100, fall cone gives slightly lower values.
3. Some standards specify slightly different cone mass, apex angle, and the penetration depth (Leroueil and Le Bihan 1996).
4. BS 1377 recommends plotting both water content and penetration on arithmetic scales. AS 1289.3.9.1 recommends plotting water content on arithmetic scale and penetration on logarithmic scale.

Datasheet:

A simple datasheet for this test, prepared in Excel, is shown in Table B10.1. The liquid limit should be reported as an integer. The other data recorded may include:

- Sample source
- Visual classification
- Percentage of material retained on 425 μm (No. 40) sieve
- Standard followed and deviations, if any
- History of the sample (e.g., natural state, air-dried, or oven-dried)

Table B10.1 Fall cone liquid limit test datasheet

Sample description:	Dark brown sandy clay with high plasticity (CH)				
Sample location:	Kirwan Hospital				
Sample No.:	TP9-3				
Date:	12-March-2005				
Tested by:	Warren O'Donnell				
Notes:					
Test no.	1	2	3	4	5
Container no.	D3	D4	D9	D13	
Mass of container (g)	12.20	12.42	12.12	12.56	
Mass of cont. + moist soil (g)	26.32	28.54	25.04	28.83	
Mass of cont. + dry soil (g)	21.23	22.63	20.21	22.65	
Mass of water (g)	5.09	5.91	4.83	6.18	
Mass of dry soil (g)	9.03	10.21	8.09	10.09	
Water content (%)	56.37	57.88	59.70	61.25	
Penetration (mm)	14.2	18.5	22.3	25.6	

Analysis:

Figure B10.2 shows the plot of water content against the penetration. The line of best fit shown in the figure is known as the *flow curve*. The water content corresponding to penetration of 25 mm is read off the plot as 59, which is the liquid limit. A blank graph sheet is given in Figure B10.3.

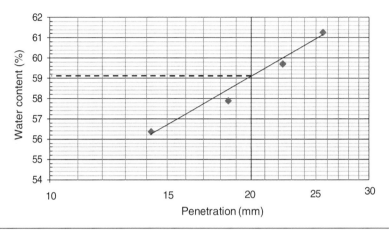

Figure B10.2 Water content vs. penetration plot

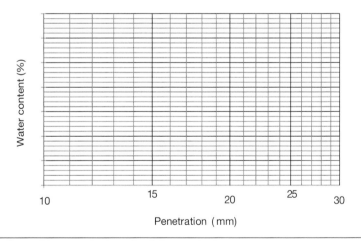

Figure B10.3 A blank semi-log graph sheet for fall cone test

Cost: US$70–US$80

B11   Plastic Limit

Objective: To determine the plastic limit of a fine grained soil

Standards: ASTM D4318
AS 1289.3.2.1
BS 1377-2

Introduction

Plastic limit, denoted by *PL* or w_p, is the lowest water content at which a fine grained soil is still plastic. At water content lower than this, the soil will be semiplastic and cannot be rolled into a thread. Plastic limit is defined as the water content at which the fine grained soil just crumbles when rolled into a 3 mm diameter thread. *Plasticity index* is defined as:

$$\text{Plasticity index } (PI) = \text{Liquid limit } (LL) - \text{Plastic limit } (PL) \qquad \text{(B11.1)}$$

Plasticity index is the measure of plasticity of a fine grained soil. Clays are plastic and silts are nonplastic.

Procedure:

1. Obtain 40 g of soil passing a 425 µm (No. 40) sieve.
2. Thoroughly mix the soil with distilled water to a consistency at which the soil can be rolled without sticking to the fingers. Shape some of this into a ball and roll on a glass plate into a thread until it crumbles.
3. If the thread does not crumble at a diameter of 3 mm, it is too wet; if it crumbles at a diameter larger than 3 mm, it is too dry. If it crumbles at a diameter of 3 mm, the clay is at its plastic limit. Collect the crumbled threads and determine the water content, which is the plastic limit. Use the 3 mm and 100 mm long metal rods for comparison (Figure B11.1).
4. If the thread does not crumble at 3 mm diameter, it is too wet. Knead between fingers to lose some moisture and try again.
5. If the thread crumbles before reaching 3 mm diameter, it is too dry. Add some water and remix the soil and knead thoroughly before rolling into a thread.
6. Take the average value of two separate determinations of the plastic limit. If the two results differ by more than the range specified by the standard, the test should be repeated. ASTM, BS, and AS allow the difference to be within 1, 0.5, and 2% water content, respectively.

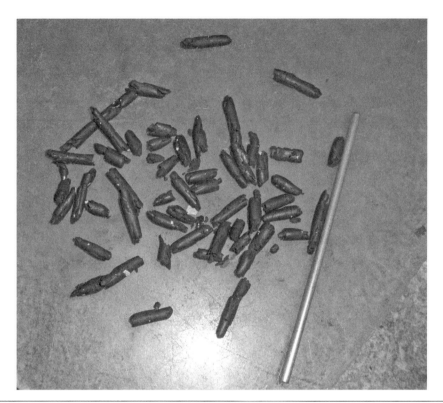

Figure B11.1 Plastic limit test (Photograph courtesy of Ms. Dhanya Gaseralingam, James Cook University, Australia)

Datasheet:

Computation of linear shrinkage is fairly straightforward, and there is no special datasheet for this test. Plastic limit should be reported as an integer. If the plastic limit cannot be determined, or if it turns out to be greater than the liquid limit, the soil should be reported as nonplastic. The other data that can be recorded are:

- Sample source
- Percentage of material retained on a 425 μm (No. 40) sieve
- Standard followed and deviations, if any
- History of the sample (e.g., natural state, air-dried, or oven-dried)

Analysis:

None

Cost: US$40–US$50

B12 Linear Shrinkage

Objective: To determine the linear shrinkage of a fine grained soil

Standards: AS 1289.3.4.1
BS 1377-2

Introduction

Linear shrinkage is an indirect measure of plasticity. For fine grained soils with low clay content, the standard liquid limit and plastic limit tests may not give reliable results. For such soils, a linear shrinkage test can be used for estimating the plasticity index as:

$$\text{Plasticity index} = 2.13 \times \text{Linear shrinkage} \qquad \text{(B12.1)}$$

The 250 mm long linear shrinkage metal mold shown in Figure B12.1 consists of a 25 mm diameter semicylindrical trough, with ends brazed on normal to the longitudinal axis of the mold. The mold is filled with soil paste at a water content close to the liquid limit and is dried in the oven. The reduction in the soil paste length, expressed as a percentage, is the linear shrinkage.

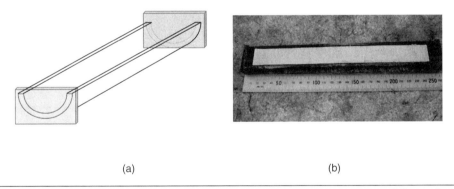

(a)   (b)

Figure B12.1 (a) Linear shrinkage mold and (b) mold with dry soil (Courtesy Warren O'Donnell, James Cook University, Australia)

Procedure:

1. Obtain 250 g of soil passing a 425 μm (No. 40) sieve.
2. Place the entire sample in a mixing bowl and add water gradually to make a thick homogeneous paste by mixing thoroughly. Leave this mixture to cure and for the moisture to equilibrate for 12 to 24 hours.

3. Add sufficient water to the cured soil and mix thoroughly to bring the consistency of the paste to the liquid limit.
4. Using a spatula or palette knife, fill the linear shrinkage mold with the soil paste. Strike off the excess, leaving a smooth flat surface level with the rim of the mold.
5. Most standards recommend drying in two stages. The first stage of drying can be at room temperature for 24 hours before the mold is transferred to the oven to complete the drying until there is no further shrinkage.

Datasheet:

The computation of linear shrinkage is fairly straightforward, and there is no special datasheet for this test. When a 250 mm long mold is used, linear shrinkage should be reported to the nearest 0.5%. When a 135 mm mold is used, it should be reported to the nearest 1%. The other data that can be recorded are:

- Sample source
- Visual classification
- Standard followed and deviations, if any

Analysis:

The linear shrinkage is defined as:

$$\text{Linear shrinkage (\%)} = \frac{\Delta L}{L} \times 100 \tag{B12.2}$$

where L = initial length of the paste (mold length) and ΔL = reduction in the length of the paste when dried.

Cost: US\$40–US\$60

B13   Compaction Test

Objective: To determine the water content versus dry density relationship of a soil or aggregate and identify the optimum water content and maximum dry density at a specific compactive effort

Standards: ASTM D698 & D1557
AS 1289.5.1.1 & 1289.5.2.1
BS 1377-4

Introduction

Compaction is the most common ground improvement technique applied to nearly all types of soils. Depending on the soil, different types of rollers are used in the field to compact the soil in 150 to 500 mm layers known as *lifts*. To simulate the field compaction in the laboratory, soil is placed in a 100 mm or 150 mm diameter metal cylindrical mold and compacted in 3 or 5 layers using a specific compactive effort. This is repeated for 5 or 6 different water contents, and the dry density of the compacted earth is plotted against the molding water content. A typical compaction curve is shown in Figure B13.1. The *maximum dry density* $\rho_{d,max}$ is achieved at a specific water content known as the *optimum water content* w_{opt}. When the compactive effort

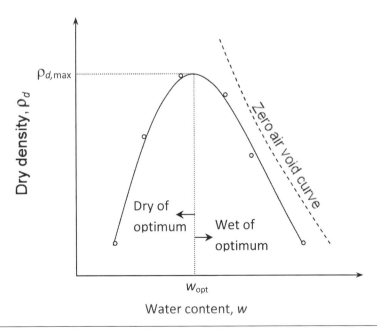

Figure B13.1 A typical compaction curve with zero air void curve

is increased, the compaction curve shifts upward and left, resulting in an increase in $\rho_{d,\max}$ and reduction in w_{opt}.

The laboratory compaction is carried out on the soil contained within a mold (Figure B13.2a) using a hammer (Figure B13.2b) that falls from a certain height. There are two compactive efforts commonly used in the laboratories: (1) *standard Proctor* and (2) *modified Proctor* compactive efforts. The hammer mass, drop height, number of blows per soil layer, number of layers, and other details are summarized in Table B13.1. There are some differences between the various standards. For most soils, compaction is carried out in a 100 mm diameter mold. When there is a significant gravel component in the soil, larger 150 mm diameter CBR molds are used. With a larger mold, and hence larger soil volume, the number of blows has to be increased as given in Table B13.1.

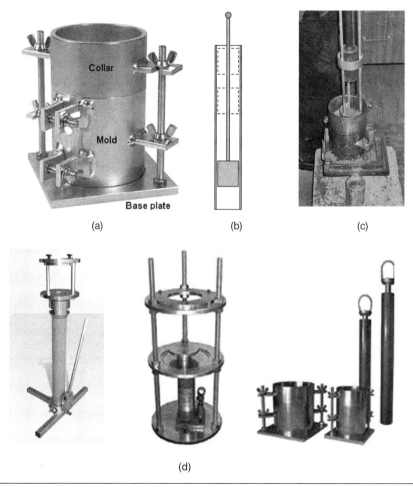

(a)   (b)   (c)

(d)

Figure B13.2 Compaction test: (a) compaction mold, collar, and base plate; (b) hammer; (c) a test in progress; and (d) sample extruders, molds, and hammers (Photograph: N. Sivakugan)

Table B13.1 Test details as per ASTM, AS, and BS

	ASTM		AS		BS	
	Standard	**Modified**	**Standard**	**Modified**	**Standard**	**Modified**
Hammer:						
Mass, kg	2.50	4.53	2.70	4.90	2.50	4.50
Drop, mm	305	457.2	300	450	300	450
Mold:						
Diameter, mm	101.6	101.6	105.0	105.0	105.0	105.0
Height, mm	116.4	116.4	115.5	115.5	115.5	115.5
Volume, cm^3	943**	943**	1000	1000	1000	1000
No. of layers	3	5	3	5	3	5
Blows per layer	25	25	25	25	27	27
Compactive effort, kN·m/m^3	600	2700	596	2703	596	2682
Alternative: Larger CBR mold						
Mold:						
Diameter, mm	152.4	152.4	152.0	152.0	152	152
Height, mm	116.4	116.4	132.5	132.5	127	127
Volume, cm^3	2124	2124	2400	2400	2305	2305
No. of layers	3	5	3	3	3	5
Blows per layer	56	56	60	100	62	62
Compactive effort, kN·m/m^3	592	2678	596	2703	594	2672

**1/30 ft^3 as originally used by R.R. Proctor (1933)

Generally, the best geotechnical characteristics of a compacted earthwork are achieved in the vicinity of the optimum water content, say ± 2% of optimum water content. Soil properties can be very sensitive to the molding water content, especially for clayey soils. Therefore, there is stringent control on most earthworks through proper specifications.

The dashed zero air void curve shown in Figure B13.1 is a theoretical curve plotted on the ρ_d-w space using the following equation with the correct value of the specific gravity G_s.

$$\rho_d = \frac{G_s \rho_w}{1 + w G_s} \qquad \text{(B13.1)}$$

All the test points must lie to the left of the zero air void curve. A point lying to the right implies that the degree of saturation is greater than 100%, which is impossible. Therefore, the theoretical zero air void curve serves as a check on the compaction test data. It is a good practice to draw this along with the compaction test data. The location of the zero air void curve is influenced significantly by the value of G_s.

Procedure:

1. *Selection of the mold based on grain sizes:* ASTM, AS, and BS have slightly different ways of deciding between the small and large compaction molds.

   - ○ ASTM recommends that compaction tests be carried out on soils where not more than 30% is coarser than 19 mm. Depending on the grain sizes present, three methods are suggested: (1) Method A using soil passing a 4.75 mm sieve; (2) Method B using soil passing a 9.5 mm sieve; and (3) Method C using soil passing a 19 mm sieve. Methods A and B use the smaller mold and Method C uses the larger CBR mold. Method A can be applied to soils where less than 25% is coarser than 4.75 mm, and Method B can be applied to soils where less than 25% is coarser than 9.5 mm. Method C is used when less than 30% is retained on a 19.0 mm sieve. If the coarser fraction is 5 to 25%, an over-size correction has to be applied. In situations where both Methods A and B meet the requirements, Method A should be used unless specified otherwise by the client.

   - ○ Australian Standards recommends that soil passing a 37.5 mm sieve be used in compaction tests. If less than 20% is coarser than 19.0 mm, only the fraction less than 19 mm is tested using the smaller mold. If there are more than 20% of 19.0 to 37.5 mm particles, but less than 20% larger than 37.5 mm, only the soil passing 37.5 mm is tested in the larger CBR mold. If there are more than 20% grains coarser than 37.5 mm, the soil is too coarse for a compaction test.

   - ○ British Standards recommends that soil passing 37.5 mm be used in a compaction test. If 95 to 100% passes a 20 mm sieve, after removing the particles larger than 20 mm, the material can be tested in the small mold. If there are significant 20 to 37.5 mm particles present, but less than 10% larger than 37.5 mm, the particles larger than 37.5 must be removed and the remainder tested in a larger CBR mold. If there are more than 30% grains coarser than 20 mm, the soil is too coarse to be tested.

2. *Optimum water content:* Before starting the test, it is desirable to have an idea about the magnitude of the expected optimum water content. This way, the test water contents can be decided so that they spread evenly about the optimum. For cohesive soils, the optimum water content (for standard Proctor compactive effort) is slightly less than the plastic limit. In sands, optimum water content is where you can squeeze out a drop of water. Plan to have five specimens where two are at water contents less than the optimum and two at water contents greater than optimum (spread by about 1 to 1.5% for crushed rocks and gravels and 2 to 3% for clays between the adjacent points).

3. *Sample preparation:* ASTM does not recommend reuse of previously compacted soils; recompacted clays clump together and give greater ρ_d values. BS does not recommend reuse of compacted soils when the grains are susceptible to crushing. The small and large molds require 2.5 kg and 6.5 kg of soil, respectively, for each water content. This

translates to approximately 15 kg to 30 kg of dry soil for the entire compaction test when the soils are not reused.

4. Mix the sub-samples at desired water contents and cure overnight.

5. Assemble the mold on to the base plate and attach the collar (Figure B13.2a). Place the mold on a rigid foundation. Place the soil into the mold and compact into 3 to 5 layers of equal thickness using the appropriate hammer (Figures B13.2b and B13.2c) and number of blows (see Table B13.1). Scarify the surface of the previous layer before adding the next layer. At the end of compaction, the last layer should extend slightly into the collar, but not more than 6 mm.

6. Remove the collar and scrape off the excess above the rim of the mold, using a metal straightedge or a knife.

7. Determine the mass of the compacted soil and the bulk density ρ_m.

8. Remove the specimen from the mold. A hydraulic jack may be useful for the extrusion.

9. Determine the water content w using a representative sample or, preferably, from the whole specimen.

10. Determine the dry density ρ_d using the following relationship:

$$\rho_d = \frac{\rho_m}{1 + w} \tag{B13.2}$$

where w is expressed as a decimal number and not as a percentage.

11. Repeat Steps 5 to 9 for all sub-samples.

Datasheet:

A simple datasheet for a compaction test prepared in Excel is shown in Table B13.2, and the compaction curve is shown in Figure B13.3. The optimum water content and the maximum dry density should be reported to the nearest 0.1% and 0.01 Mg/m^3, respectively. Note that 1 Mg/m^3 = 1 tonne/m^3 = 1 g/cm^3. The other data that can be recorded are:

- Sample source and description of the material
- As received water content or hygroscopic water content
- History of the sample (e.g., natural state, air-dried, or oven-dried)
- Grain size data and method used (A, B, or C)
- Percentage of material retained on a 425 μm (No. 40) sieve
- Sample preparation methods
- Compactive effort (e.g., modified Proctor)
- Specific gravity of the grains to nearest 0.01 and the method of determination
- Standard followed and deviations, if any
- Compaction curve, zero air void curve, and the values of the optimum water content and the maximum dry density (or unit weight)

Table B13.2 Compaction test datasheet

Sample description:	Gravelly, sandy clay of high plasticity					
Sample location:	Townsville port					
Average specific gravity	2.80	Sample no.	TPA-43			
Grading		Date:	12th June 2010			
< 4.75 mm fraction	78	Tested by:	Warren O'Donnell			
4.75-9.5 mm fraction	15	Comments:				
9.5-19.0 mm fraction	7					
19.0-37.5 mm fraction	0					
> 37.5 mm fraction	0					

Compactive effort:	Standard Proctor				
Bulk density ρ_m:	Test 1	Test 2	Test 3	Test 4	Test 5
Mold number	A1	A1	A1	A1	A1
Mold volume (cm³)	943	943	943	943	943
Mass of mold (g)	1942.1	1942.1	1942.1	1942.1	1942.1
Mass of mold + wet soil (g)	3732.2	3843.2	3932.4	3913.8	3867.4
Bulk density ρ_m: (Mg/m³)	1.90	2.02	2.11	2.09	2.04
Water content w:					
Tin (or tray) number	42	21	16	17	19
Mass of tin (g)	32.23	31.46	31.89	32.92	32.12
Mass of tin + wet soil (g)	123.51	132.52	129.32	132.73	122.81
Mass of tin + dry soil (g)	112.90	119.30	114.98	116.42	106.79
Water content (%)	13.15	15.05	17.26	19.53	21.45
Dry density, ρ_d (Mg/m³)	1.68	1.75	1.80	1.75	1.68
ρ_d (Mg/m³) for zero air void curve	2.05	1.97	1.89	1.81	1.75

Analysis:

Sometimes it is useful to plot the theoretical degree of saturation (S) or air content (a) contours. The degree of saturation is defined as the percentage of the void volume that is occupied by water and can vary in the range of 0 (dry soil) to 100 (saturated soil). Air content is the percentage of the total volume that is occupied by air. Their contours can be plotted using Equations B13.3 and B13.4.

$$\rho_d = \frac{G_s \rho_w}{1 + \dfrac{wG_s}{S}} \tag{B13.3}$$

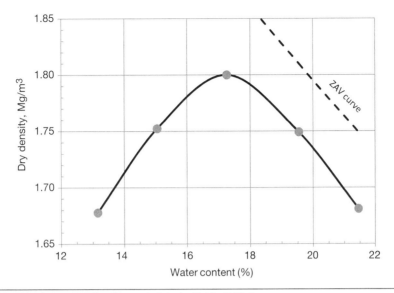

Figure B13.3 Compaction curve for the test data in Table B13.2

$$\rho_d = \frac{G_s(1-a)\rho_w}{1+wG_s}$$ (B13.4)

where G_s is the specific gravity of the soil grains. These theoretical contours enable better visualization of the air void volume in the compacted soil when the test points are plotted.

Cost: US$180–US$200

B14 Maximum and Minimum Densities of a Granular Soil

Objective: To determine the maximum and minimum void ratios (or densities) of a granular soil

Standards: ASTM D4253 & D4254
AS 1289.5.5.1
BS 1377-4

Introduction

Granular soils are relatively porous and well drained. As a result, their compaction characteristics are less affected by the water content, and it is not common to see the classical compaction curve seen in cohesive soils.

The grain size distribution plays a key role in the geotechnical behavior of coarse grained soils. For the same grain size distribution, the soils can exhibit quite different strength and stiffness characteristics, depending on how closely the grains are packed. Generally, the strength (i.e., friction angle) and stiffness (e.g., Young's modulus) increase with the packing density. The density of packing at the void ratio e is quantified through a term known as *relative density D_r* that lies within the range of 0 to 100%. It is defined as:

$$D_r(\%) = \frac{e_{max} - e}{e_{max} - e_{min}} \times 100 \qquad (B14.1)$$

where e_{max} and e_{min} are the *maximum and minimum void ratios* that can be achieved with the same granular soil. They are simply the upper and lower bounds of the void ratios at which the soil can exist, and thus define a range ($e_{min} \le e \le e_{max}$). The soil cannot lie at a void ratio outside this range. When $e = e_{max}$, $D_r = 0\%$ and when $e = e_{min}$, $D_r = 100\%$. Relative density D_r is also known as *density index I_D*. The term relative density is meaningless in cohesive soils.

Dry density and void ratio are related by:

$$\rho_d = \frac{G_s \rho_w}{1 + e} \qquad (B14.2)$$

where G_s is the specific gravity of the soil grains. In terms of dry densities, Equation B14.1 can be written as:

$$D_r(\%) = \frac{\rho_d - \rho_{d,min}}{\rho_{d,max} - \rho_{d,min}} \times \frac{\rho_{d,max}}{\rho_d} \times 100 \qquad (B14.3)$$

where $\rho_{d,max}$ and $\rho_{d,min}$ are the maximum and minimum possible dry densities corresponding to D_r of 100% and 0% respectively, where the void ratios are e_{min} and e_{max} respectively (see Figure

B14.1). ρ_d is the current dry density, corresponding to void ratio e at which the relative density is being determined. Note that maximum void ratio corresponds to minimum density and vice versa (see Equation B14.2). The objective of this test is to determine e_{max} and e_{min} so that the relative density can be defined at a specific void ratio. This requires two separate tests that are described below.

Granular soils are often classified on the basis of relative density as shown in Figure B14.1. Some typical values of e_{max} and e_{min} are given in Table B14.1. Hilf (1975) summarized more values, covering a wide range of granular soils. In uniformly graded soils, the range $e_{max} - e_{min}$ is relatively small and they are difficult to compact. Terzaghi (1925) used $(e_{max} - e_{min})/e_{min}$ as a measure of compatibility for granular soils, with larger values suggesting easier compaction.

Table 14.1 Typical values of e_{max}, e_{min}

| Soil | G_s | Loosest (D_r = 0%) | | Densest (D_r = 100%) | |
		e_{max}	$\rho_{d,min}$ (Mg/m³)	e_{min}	$\rho_{d,max}$ (Mg/m³)
Standard Ottawa sand**	2.65	0.80	1.48	0.50	1.76
Clean uniform sand**	2.65	1.00	1.33	0.40	1.89
Uniform inorganic silt**	2.67	1.10	1.27	0.40	1.91
Silty sand**	2.65	0.90	1.39	0.30	2.04
Fine to coarse sand**	2.65	0.95	1.36	0.20	2.21
Micaceous sand**	2.68	1.20	1.22	0.40	1.91
Silty sand and gravel**	2.65	0.85	1.43	0.14	2.32
Hydraulic minefill#	2.79	0.94	1.44	0.45	1.92
Hydraulic minefill#	3.42	1.04	1.68	0.53	2.24
Hydraulic minefill#	4.37	1.05	2.13	0.67	2.62

**Lambe and Whitman (1979), #Rankine et al. (1996)

Procedure:

Maximum density (wet placement):

The granular soil is brought to the densest state through saturation and vibration. The vertical vibration is provided by an electromagnetic, eccentric or cam-driven vibrating table, vibrating

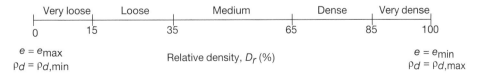

Figure B14.1 Classifying granular soils based on relative density

at 50-60 Hz. For uniform spherical grains, it can be shown theoretically that the minimum void ratio is 0.350.

1. *Soil:* The test is applicable for soils containing less than 15% fines and where the fines are cohesionless and free-draining. It is also necessary that the entire soil is finer than 75 mm.
2. *Mold:* ASTM suggests two different molds (2,830 cm^3 and 14,200 cm^3), depending on the grain sizes. BS recommends 1000 cm^3 mold for sandy soils and the CBR mold for gravelly soils. AS permits the use of 1,000 cm^3 compaction mold for soils that are pre-dominantly sands.
3. *Sample preparation:* Add water to the granular soil sample in a mixing bowl and soak for an hour. The amount of water added should be such that there is no free water and that the specimen is saturated by the vibratory densification process.
4. *Mold volume calibration:* Determine the mold volume from the diameter and height measured to 0.1 mm or by filling with water. It should be within ± 1.5% of the nominal volume. It is recommended that both methods be used to determine the volume and if the difference exceeds 0.5% of the nominal volume, the calibration can be repeated.
5. Place the mold (with collar attached) on the vertically vibratory table.
6. Turn on the vibrating table and slowly fill the mold with wet granular soil, using a scoop or shovel. Overfill the mold by 20 mm, allowing for settlement. Continue the vibration for five more minutes and remove any excess water at the surface.
7. Place the surcharge plate that typically applies about 14 kPa on top of the granular soil and vibrate for 10 minutes.
8. Turn off the vibration and remove the surcharge plate and the mold assembly. Level off the top of the surface with a straightedge.
9. Determine the mass of the saturated sand.
10. Place the entire specimen in a tray and determine the water content.

Notes:

1. ASTM suggests an alternate dry placement method (Method 1A or 2A) that uses the same vertically vibrating table. The vibrating table can be electromagnetic (Method 1A and 1B) or eccentric or cam-driven (Method 2A or 2B).
2. The dry placement method is quicker and is often preferred. Wet placement can produce higher densities in some soils. When in doubt, wet placement is suggested.

Minimum density (dry placement):

The objective here is to place the granular soil at its loosest possible state (e_{max} and $\rho_{d,min}$) and this is achieved by providing the maximum possible hindrance during deposition or by fast deposition. There are three different ways suggested in ASTM D4254: (1) funnel method—using a funnel and tube to deposit the dry soil in the mold, keeping the drop height small; (2) double tube method—placing a tube filled with dry soil in the mold and removing the tube, thus allowing the soil to fill the mold; and (3) depositing the soil by inverting a graduated cylinder. The first method is preferred. For uniform spherical grains, it can be shown theoretically that the maximum void ratio is 0.91.

1. *Soil:* The test is recommended for soils having less than 15% fines, where the fines are cohesionless and free-draining. Use oven-dried soils in the test.
2. *Mold:* As before.
3. Place the granular soil into the mold using a funnel and tube (or by any alternate method), allowing free fall of about 10 mm, and determine the mass of the soil required to fill the mold and, hence, the density $\rho_{d,min}$.

Datasheet:

A simple datasheet for this test, prepared in Excel, is shown in Table B14.2. The maximum and minimum dry densities should be reported to the nearest 0.01 Mg/m^3. If expressed as unit weight, it should be to the nearest 0.01 kN/m^3. Note that 1 Mg/m^3 = 1 tonne/m^3 = 1 g/cm^3. The other data recorded may include the following:

- Sample source and description
- Size of the mold
- Details of the vibrating table (e.g., model, frequency) used in the maximum density test
- Surcharge mass (in the maximum density test)
- Abnormalities such as loss of material, segregation and tilt of the base plate
- Standard followed and deviations, if any

Analysis:

As shown in Table B14.2, the computations are fairly straightforward. It simply requires the calculation of densities from the mass and volume of a cylindrical sample. From the maximum and minimum densities, and the right value of specific gravity G_s, the minimum and maximum void ratios can be determined using Equation B14.2. In the absence of test data for G_s, it can be assumed as 2.65 for most granular soils.

Table 14.2 Maximum and minimum dry density datasheet

Sample description:	Well graded sand				
Sample location:	Dairy Framers Stadium				
Sample No.	TPA-43				
Date	18th July 2004	Tested by: Rudd Rankine			
Comments:	As per AS 1289.5.5.1				

Maximum dry density, $\rho_{d, max}$:

Vibrating table details: Syntron Magnetic vibrator V-51-D1; 50 Hz

	Test 1	Test 2	Test 3	Test 4	Test 5
Soaking time	1 hr	1 hr	1 hr		
Surcharge mass (kg)	5.0	5.0	5.0		
Vibration time pre-surcharge	9 min	10 min	9 min		
Vibration time post-surcharge	10 min	10 min	10 min		
Mass of mold (g)	1942.6	1942.6	1942.6		
Mass of mold + wet specimen (g)	3962.6	3952.7	3964.2		
Volume of the mold (cm^3)	943.0	943.0	943.0		
Bulk density, ρ_m (Mg/m^3)	2.14	2.13	2.14		
Mass of tray (g)	235.6	235.6	235.6		
Mass of tray + wet specimen (g)	2251.4	2238.6	2248.5		
Mass of tray + dry specimen (g)	1941.3	1934.3	1930.1		
Water content (%)	18.18	17.91	18.79		
Maximum dry density, ρ_d (Mg/m^3)	1.813	1.808	1.805		
Minimum dry density, $\rho_{d, min}$:	Test 1	Test 2	Test 3	Test 4	Test 5
Mass of mold (g)	1942.6	1942.6	1942.6		
Mass of mold + dry soil (g)	3245.7	3247.3	3244.1		
Volume of the mold (cm^3)	943.0	943.0	943.0		
Minimum dry density, ρ_d (Mg/m^3)	1.382	1.384	1.380		

Cost: US$250–US$300

B15   Field Density Test

Objective: To determine the in-place density of a soil

Standards: ASTM D1556, D2167, D2937 & D5030
AS 1289.5.3.1 and 1289.5.3.2
BS 1377-9

Introduction

The quality of compaction in an earthwork is assessed periodically (e.g., one test per every 1000 m^3) through some control tests where the dry density and the water content of the compacted soil are measured to check whether it meets the specifications. These tests include:

- Sand cone method
- Rubber balloon method
- Core cutter method
- Water replacement method
- Nuclear method

The *sand cone method* (Figure B15.1a) requires a hole dug in the ground from which the soil is removed and weighed. The volume of the hole is determined by filling it with uniform sand (e.g., Ottawa sand) of a known density. From the mass of sand required to fill the hole, the volume can be calculated. From the mass of the soil removed from the hole, and the volume determined from the sand cone apparatus, the bulk density can be calculated.

The water content of the soil is also determined from a sample, and the dry density is calculated from:

$$\rho_d = \frac{\rho_m}{1 + w} \tag{B15.1}$$

where water content w is a decimal number not a percentage.

The *rubber balloon method* (Figure B15.1b) is similar to the sand cone method except that the volume of the hole is determined by lining it with a rubber balloon and filling it with water from a calibrated vessel.

The *core cutter method* is applicable in cohesive soils where a thin-walled sampling tube is driven into the soil. From the mass, volume, and the water content of the cylindrical soil specimen, the dry density is determined.

The *water replacement method* is carried out to assess the average density or unit weight of test pits of large volumes. The volume of the pit is determined by lining the pit and filling it with water.

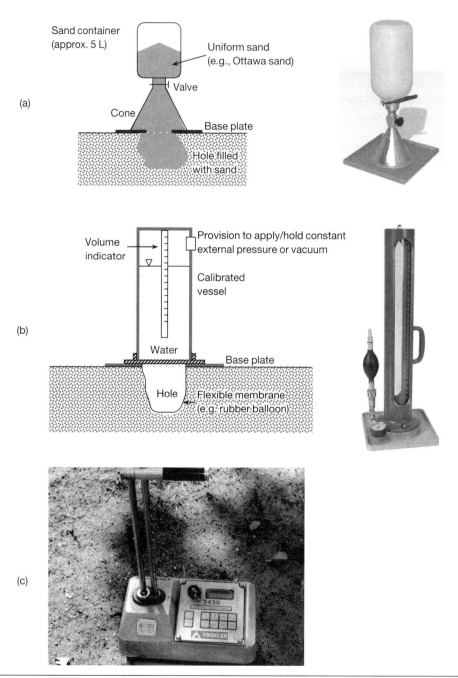

Figure B15.1 Field density measurement devices: (a) sand cone; (b) rubber balloon; and (c) nuclear density meter (Photograph: N. Sivakugan)

Nuclear density meters (Figure 15.1c) are commonly used for the quick determination of dry density and water content of compacted earthwork. It is a nondestructive test.

Procedure:

Sand cone method (ASTM D1556):

1. *Test location:* Select a test location that is representative of the area being tested and prepare the surface as a level plane. Seat the base plate on the plane surface.
2. *Test hole:* The test hole should be as large as practicable. The minimum size recommended by ASTM is given in Table B15.1. Dig the test hole through the center of the base plate.
3. Place the excavated soil in a moisture tight container for later determination of the mass of the soil (m_1) at its in-place water content.
4. Take a sample (later in the laboratory) from this soil and determine the water content (w) to the nearest 1%.
5. Invert the sand cone apparatus and seat the sand cone funnel on the base plate, as shown in Figure B1a, and open the valve allowing the sand to fill the hole, funnel, and the base plate. When there is no more sand flowing, close the valve and determine the mass of the apparatus with the remaining sand (m_2).
6. *Calibration:* Through a calibration exercise, determine the density of the sand used in the sand cone (ρ_{sand}) and the mass of sand required to fill the cone and base plate (m_3).
7. *Sand:* The sand used in the test should be clean, uniform sand with a coefficient of uniformity C_u ($= D_{60}/D_{10}$) less than 2, and a maximum grain size less than 2 mm so that there will be minimum volume change during handling, less segregation, and no clogging of the valve, ensuring free flow. *Ottawa sand* and *silica sand* are two commonly used sands that suit this purpose.

Rubber balloon method (ASTM D2167):

1. *Test location:* Select a test location that is representative of the area being tested and prepare the surface as a level plane. Seat the base plate on the plane surface.

Table B15.1 Minimum test hole volume for sand cone and rubber balloon methods

Size of the largest grain	Minimum volume (cm³)
12.7 mm (½ in.)	1415
25.4 mm (1 in.)	2125
37.5 mm (1½ in.)	2830

2. *Test hole:* The test hole should be as large as practicable. The minimum size recommended by ASTM is given in Table B15.1. Dig the test hole through the center of the base plate.
3. Place the excavated soil in a moisture tight container and, later, record the mass of the soil (m_4) at its in-place water content.
4. Take a sample (later in the laboratory) from this soil and determine the water content (w) to the nearest 1%.
5. Place the apparatus over the base plate and determine the volume (V) of water required to fill the hole that is lined with the rubber balloon contained in the apparatus.

Core cutter method (ASTM D2937):

1. *Core cutter:* ASTM and BS recommend approximately 100 mm diameter and 130 mm long core cutters with thin walls such that the area ratio A_R is preferably less than 15%. Area ratio A_R is defined as:

$$A_R(\%) = \frac{D_o^2 - D_i^2}{D_i^2} \times 100 \tag{A.2}$$

A drive head can be used to drive the core cutter into the ground.
2. Determine the mass of the core cutter (m_5).
3. Drive the core cutter into the ground. Use the drive head, drop weights, and some effort as appropriate. Stop driving when the top of the core cutter is about 13 mm below the original ground level.
4. It may be necessary to excavate around the core cutter to remove it without any disturbance.
5. Trim any excess soil from the ends and the wall and determine the mass of the core cutter with the soil (m_6).
6. Volume of the core cutter (V^{*}) must be determined to the nearest 0.01 cm^3 from the dimensions.
7. Take a sample from the core and determine the water content (w).

Datasheet:

The calculations are rather straightforward and, hence, no datasheets are included for this test. The following information can be included in the test report:

- Test location
- Test method and the apparatus details
- Visual description of the soil removed from the hole
- In-place bulk and dry densities
- In-place bulk and dry unit weights

- In-place water content
- Any deviations from the standards

In the sand cone test, the bulk density of the sand may be included. In the case of the core cutter method, include the dimensions and volume of the core cutter, the methods used in driving the core cutter into the ground, and any comments on possible soil disturbance.

Analysis:

Sand cone method:

$$\text{Mass of the soil removed from the hole} = m_1$$

$$\text{Volume of the hole, } V = \frac{m_2 - m_3}{\rho_{sand}}$$

$$\therefore \text{Bulk density, } \rho_m = \frac{m_1}{m_2 - m_3} \times \rho_{sand}$$

Rubber balloon method:

The calculations are similar to the above and $\rho_m = \frac{m_4}{V}$.

Core cutter method:

Bulk density, $\rho_m = \frac{m_6 - m_5}{V*}$

Cost: US$125–US$150

B16   Hydraulic Conductivity of a Coarse Grained Soil

Objective: To determine the hydraulic conductivity of a coarse grained soil

Standards: ASTM D2434
   AS 1289.6.7.1
   BS 1377-6

Introduction

Water flows from higher head to lower head, where the head loss per unit length is a dimensionless quantity known as *hydraulic gradient* (i). Darcy's law for laminar flow through soils states that the *discharge velocity* (v) is proportional to the hydraulic gradient, and it can be written as:

$$v = k\,i \tag{B16.1}$$

where k is known as *hydraulic conductivity* or *permeability*, which has a unit of velocity. It is a measure of how easily water can flow through soils. It is an important parameter in geotechnical problems involving seepage or dewatering. Lambe (1951) suggested permeability of 10^{-4} cm/s as the borderline between pervious and poorly drained soils. Values significantly less (e.g., 10^{-7} cm/s for compacted clay liners) are suggested currently for ensuring that the soil is impervious. Typical values of permeability of coarse grained soils are given in Table B16.1. The soil classification based on permeability values is shown in Table B16.2.

For clean sands, with $D_{10} = 0.1 - 3.0$ mm, permeability can be crudely estimated as (Hazen 1911):

$$k\,(\text{cm/s}) \approx D_{10}^{2}\,(\text{mm}) \tag{B16.2}$$

In coarse grained soils, permeability k is a function of the void ratio e and is often proportional to e^{2}, $e^{2}/(1 + e)$, or $e^{3}/(1 + e)$. In general, e versus $\log k$ is approximately a straight line for any soil (Lambe and Whitman 1979).

Table B16.1 Typical values of permeability

Soil	k (cm/s)
Fine sand	$10^{-3} - 10^{-2}$
Clean sands, sand/gravel mix	$10^{-3} - 10^{0}$
Coarse sand	$10^{-2} - 10^{0}$
Clean gravel	$10^{0} - 10^{2}$
#Ottawa sand ($e \approx 0.63$)	6.5×10^{-3}
#Beach sand ($e \approx 0.75$)	1.5×10^{-1}
#Silty sand ($e \approx 0.27$)	2.0×10^{-8}

Lambe and Whitman (1979)

Table B16.2 Classification based on permeability (after Terzaghi and Peck 1967)

Degree of permeability	k (cm/s)
Practically impermeable	$< 10^{-7}$
Very low	$10^{-7} - 10^{-5}$
Low	$10^{-5} - 10^{-3}$
Medium	$10^{-3} - 10^{-1}$
High	$> 10^{-1}$

In the laboratory, permeability of an undisturbed or reconstituted soil specimen can be determined by a *constant head* or *falling head permeability test*. The nature of the test is such that the constant head test is commonly carried out for coarse grained soils and the falling head test is performed for fine grained soils. The schematic diagram of a simple setup for carrying out a constant head permeability test is shown in Figure B16.1. This is an inexpensive apparatus that can easily be developed in the laboratory.

In coarse grained soils, it is difficult to get undisturbed samples. The constant head test described here is mainly for reconstituted specimens where the granular soils are packed at desired densities.

Reynolds' number (R) is used to define the boundaries between laminar and turbulent flows. It is defined as:

$$R = \frac{vd\rho_w}{\mu_w} \qquad \text{(B16.3)}$$

where v = discharge velocity, d = pipe diameter, ρ_w = density of water, and μ_w = dynamic viscosity of water. In pipe flows, R can be as high as 1000 while the flow is still laminar. However, in flow through soils, the geometry of the pore channels can be irregular, and it is difficult to define the pore channel diameter for Equation B16.3. Harr (1977) suggests taking d as the average grain diameter. Muskat (1937) reported that flow through soils change from laminar to turbulent when the Reynolds' number is in the range of 1 to 12. Conservatively, $R = 1$ is taken

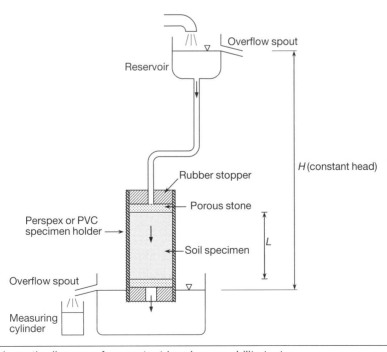

Figure B16.1 Schematic diagram of a constant head permeability test

as the upper limit to ensure the flow is laminar. In gravelly soils, the discharge velocity can be high, resulting in turbulent flow that jeopardizes the validity of Darcy's law. Casagrande (1938; see Holtz and Kovacs 1981) suggested $k = 1$ cm/s as the upper limit of permeability to ensure laminar flow. This separates clean gravels from clean sands (see Table B16.1).

Procedure:

1. *Soil specimen:* ASTM D2434 recommends the constant head test for granular soils with less than 10% fines. The diameter of the specimen should be 75 mm or more, and at least 8 (fine granular soil) to 12 (coarse granular soil) times the maximum grain size. The length across which the head loss is being measured should be greater than the diameter.
2. Prepare the specimen within the specimen holder using known mass (m_1) of dry granular soils (see Figure B16.1). Measure the length L and diameter D of the specimen. The dry density of the specimen can be computed from these. With porous stones and rubber stoppers at both ends of the specimen, place the permeameter in a container with an overflow weir filled with water and connect the top of the permeameter to the head water reservoir as shown in Figure B16.1.
3. Let the water flow through the specimen under a constant total head of H (based on the datum at the tail water level). Allow 10 to 15 minutes for the specimen to saturate and flush out any entrapped air pockets.
4. Once the flow rate is steady, collect the water flowing out of the specimen Q (cm³) in a measuring cylinder over a duration t (s) such that there is sufficient water collected for calculating the flow rate.
5. Repeat Step 4 three times.
6. Change the head H and repeat Steps 4 and 5 two more times.
7. Record the temperature of the water.

Note:

1. To ensure laminar flow, ASTM D2434 suggests a hydraulic gradient (H/L) of 0.2 to 0.3 for loose or coarser samples and a slightly higher gradient of 0.3 to 0.5 for dense or finer samples.

Datasheet:

A simple datasheet for the constant head permeability test on coarse grained soils is given in Table B16.3. The following information can be included in the test report:

*   Sample location and other pertinent information
*   Grain size analysis data, maximum grain size, any oversized grains discarded
*   Dry unit weight and void ratio of the specimen
*   Permeability (k_{20} and k_T) in m/s or cm/s

Table B16.3 Constant head permeability datasheet

Sample number: TGH_TP12

Sample description: Fine, silty sand with negligible gravels

Specific gravity (if known or estimated): 2.65

Tested by: Kirralee Rankine

Date: 20th May 2004

Specimen length L (mm)	125.6	Notes: Flushed water for 15 minutes to saturate the specimen.
Specimen diameter D (mm)	75.5	
Cross sectional area A (cm²)	44.77	

Test no.	1	2	3	4	5	6
Water collected Q (mL)	685	920	650	685	725	770
Duration t (s)	60	90	60	60	60	60
Head H (cm)	55	55	60	60	70	70
Temperature T (°C)	26	26	26	26	26	26
k_T (cm/s)	0.006	0.005	0.005	0.005	0.005	0.005
μ_T/μ_{20}	0.867	0.867	0.867	0.867	0.867	0.867
k_{20} (cm/s)	0.0050	0.0045	0.0044	0.0046	0.0042	0.0045

Analysis:

The hydraulic gradient i during the flow is given by H/L:

$$\therefore v = ki = k\frac{H}{L} \tag{B16.4}$$

The cross-sectional area A of the specimen is given by $\pi D^2/4$. If a quantity of water Q is collected in time t, the discharge velocity is given by:

$$v = \frac{Q}{At} \tag{B16.5}$$

From Equations B16.4 and B16.5:

$$k = \frac{QL}{HAt} \tag{B16.6}$$

The permeability can vary with the viscosity of water that varies with the temperature. Generally, permeability is reported at a standard temperature of 20°C. This can be determined from:

$$k_{20°C} = k_{T°C}\frac{\mu_{T°C}}{\mu_{20°C}} \tag{B16.7}$$

where $\mu_{20°C}$ = dynamic viscosity of water at 20°C and $\mu_{T°C}$ = dynamic viscosity of water at T°C. The dynamic viscosity of water at different temperatures is given in Table B16.4. ASTM D5084 suggests that $\mu_{T°C}/\mu_{20°C}$ can be approximated as:

$$\frac{\mu_{T°C}}{\mu_{20°C}} = \frac{2.2902\,(0.9842)^{T}}{T^{0.1702}} \tag{B16.8}$$

Table B16.4 Dynamic viscosity[#] of water (after Lambe 1951)

°C	0	1	2	3	4	5	6	7	8	9
0	17.94	17.32	16.74	16.19	15.68	15.19	14.73	14.29	13.87	13.48
10	13.10	12.74	12.39	12.06	11.75	11.45	11.16	10.88	10.60	10.34
20	10.09	9.84	9.61	9.38	9.16	8.95	8.75	8.55	8.36	8.18
30	8.00	7.38	7.67	7.51	7.36	7.31	7.06	6.92	6.79	6.66
40	6.54	6.42	6.30	6.18	6.08	5.97	5.87	5.77	5.68	5.58
50	5.29	5.40	5.32	5.24	5.15	5.07	4.99	4.92	4.84	4.77
60	4.70	4.63	4.56	4.50	4.43	4.37	4.31	4.24	4.19	4.13

[#]in millipoise
Note: 1 Poise = 0.1 Pa·s

Cost: US$270–US$300

B17 Hydraulic Conductivity of a Fine Grained Soil

Objective: To determine the hydraulic conductivity of a fine grained soil

Standards: AS 1289.6.7.2
  BS 1377-6

Introduction

The constant head permeability test will not produce a measurable flow rate in fine grained soils where the falling head permeability test is preferable. This test can be carried out on soils with permeability as low as 10^{-6} cm/s. For soils with very low permeability, it is recommended to carry out a constant head method using a triaxial test setup or a flexible wall permeameter (AS 1289.6.7.3). A schematic diagram of the falling head permeability test setup is shown in Figure B17.1. The head H, measured from the tail water level, is allowed to drop from the initial value of H_1 at the beginning of the test to the final value of H_2 at the end of the test. The time taken for this drop is recorded as t^*. From these measurements, and the cross-sectional areas of the specimen and the burette, permeability can be computed. There is no ASTM standard currently available for the falling head permeability test on a fine grained soil.

Procedure:

1. *Soil specimen:* The test can be carried out on remolded and compacted specimens or intact specimens collected from the site.
2. Prepare the specimen within the specimen holder, if possible. Otherwise, place the specimen prepared elsewhere (e.g., compaction mold or intact specimen from field) into the specimen holder with a water tight seal between the specimen and wall of the specimen holder. This seal can be in the form of wax or bentonite. Measure the specimen length L and diameter D and compute the cross-sectional area A.
3. Measure (if not known yet) the diameter of the standpipe and compute the cross-sectional area a.
4. With porous stones and rubber stoppers at both ends of the specimen, place the permeameter in a container with an overflow weir filled with water and connect the top of the permeameter to the vertical standpipe or burette as shown in Figure B17.1.
5. Unlike in coarse grained soils, it may require considerable effort to saturate the soil specimen. A high vacuum (e.g., 90 kPa) can be applied through a vacuum pump for several hours to remove the air pockets. Soaking the specimen in water for a few days may help in saturating the specimen. At the end, allow water to run through the specimen a few times to flush out any remaining air bubbles.

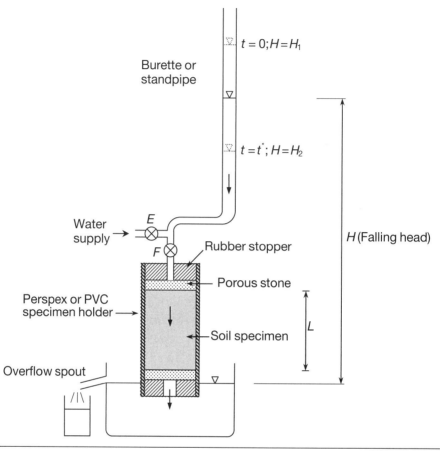

Figure B17.1 Schematic diagram of a falling head permeability test setup

6. The standpipe can be filled from water supplied through valve E while valve F is closed.
7. Raise the water level in the standpipe to $H = H_1$ and close valve E.
8. Open valve F and start the stopwatch, allowing the water level to drop. Note the time t^* taken for the water level to drop to $H = H_2$.
9. Refill the standpipe and repeat Steps 7 and 8 a few times.
10. Record the temperature of the water.

Datasheet:

A simple datasheet for a constant head permeability test on coarse grained soils is given in Table B17.1. The following information can be included in the test report:

- Remolded and compacted or intact specimen
- Sealant provided between the specimen and the permeameter wall

Table B17.1 Datasheet for falling head permeability test

Sample number: TGH_TP8

Sample description: Remolded and compacted sandy clay

Specific gravity (if known or estimated): 2.70

Tested by: Kirralee Rankine

Date: 2 July 2003

Specimen length L (mm)	125.6	Notes: : Applied vacuum for 10 hours and soaked for 72 hours.				
Specimen diameter D (mm)	75.5					
Specimen x-section area A (cm^2)	44.77					
Standpipe x-section area, a (cm^2)	0.0804					
Test no.	1	2	3	4	5	6
Initial height H_1 (mm)	795	867	840	920	880	
Final height H_2 (mm)	621	613	680	745	690	
Test duration t^* (s)	300	450	300	300	300	
Temperature T(°C)	26	26	26	26	26	
k_T (cm/s)	1.9E-05	1.7E-05	1.6E-05	1.6E-05	1.8E-05	
μ_T/μ_{20}	0.867	0.867	0.867	0.867	0.867	
k_{20} (cm/s)	1.6E-05	1.5E-05	1.4E-05	1.4E-05	1.6E-05	

- Method of saturation
- Dry unit weight and void ratio of the specimen
- Permeability (k_{20} and k_T) in m/s or cm/s

Analysis:

During the test, the head falls from H_1 to H_2 over a period t^*. The hydraulic gradient i during the flow at time t is given by H/L:

$$\therefore v = ki = k\frac{H}{L} \tag{B17.1}$$

The cross-sectional area A of the specimen is given by $\pi D^2/4$. The flow rate through the specimen is given by:

$$\dot{Q} = k\frac{H}{L}A \tag{B17.2}$$

The flow rate through the burette is given by:

$$\dot{Q} = -a\frac{dH}{dt} \tag{B17.3}$$

where a is the cross-sectional area of the burette. Equating the above two expressions for the flow rate:

$$\int_{t=0}^{t^*} dt = -\frac{aL}{Ak} \int_{H=H_2}^{H_2} \frac{dH}{H}$$

(B17.4)

Therefore, k can be expressed as:

$$k = \frac{aL}{At^*} \ln \frac{H_1}{H_2}$$

(B17.5)

The temperature correction for permeability is the same as described in the previous test (Test B16: Constant head permeability test).

Cost: US$300–US$400

B18   One-dimensional Consolidation by Incremental Loading

Objective: To determine the consolidation characteristics of a soil by incremental loading

Standards: ASTM D2435
AS 1289.6.6.1

Introduction

Consolidation is a time-dependent mechanical process where water is squeezed out of saturated soil through the application of external loads. For example, when a load is applied at ground level, the underlying saturated clay undergoes consolidation, a process that can take several years to complete. In granular soils, such as sands and gravels, this process is nearly instantaneous. Under a uniform surface load, spread over a large area, it can be assumed the consolidation is *one-dimensional*. Here, the drainage and strains are vertical only and, hence, one-dimensional. Terzaghi's (1925) one-dimensional consolidation theory suggests that the governing differential equation for the pore water pressure dissipation over time is given by:

$$\frac{\partial u}{\partial t} = c_v \frac{\partial^2 u}{\partial z^2} \tag{B18.1}$$

where u is excess pore pressure at the point that is at depth z from the top of the consolidating layer, t is time since loading, and c_v is the *coefficient of consolidation*. c_v is given by:

$$c_v = \frac{k}{m_v \gamma_w} \tag{B18.2}$$

where k = permeability and m_v = coefficient of volume compressibility, defined as the volumetric strain per unit increase in normal stress. c_v can vary from less than 1 m²/year for low permeability clays to as high as 1000 m²/year for sandy clays of very high permeability. Low c_v values are often associated with soils having a high liquid limit that take a long time to consolidate.

Consolidation tests are commonly carried out on undisturbed specimens placed in a rigid metal ring, known as *oedometer*, with a porous stone at the top and bottom of the sample to facilitate drainage. The rigid metal ring ensures that the drainage and strains are vertical only. This assembly is placed in a container filled with water (Figure B18.1a, b, c) that is placed in a loading frame (Figure B18.1d) where the pressure increments are applied, allowing consolidation to take place. Generally each pressure increment is applied over a 24-hour period during which consolidation would be mostly completed. During the consolidation process, the reduction in the sample thickness over time is monitored.

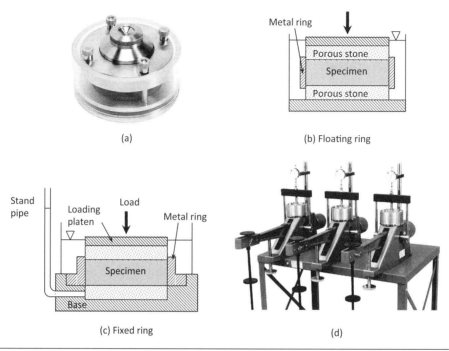

Figure B18.1 Consolidation test apparatus: (a) oedometer; (b) floating ring; (c) fixed ring; and (d) oedometers on loading frame (Photograph: N. Sivakugan)

From these data, the following parameters can be determined for every pressure increment:

- Coefficient of consolidation c_v
- Coefficient of volume compressibility m_v
- Hydraulic conductivity k
- Coefficient of secondary compression C_α

Generally, 6 to 8 successive pressure increments are applied, and, finally, the applied pressure is unloaded in 2 to 3 decrements, still recording the variation in the sample thickness over time. Thus, the complete consolidation test can take more than a week.

The coefficient of volume compressibility m_v is defined as the volumetric strain per unit increase in the normal effective stress. In one-dimensional consolidation, where there is no lateral strain, it can be shown that:

$$m_v = \frac{\Delta H}{H_0} \qquad \text{(B18.3)}$$

where H_0 = thickness of the specimen at the start of the pressure increment and ΔH = change in thickness due to consolidation. Once the consolidation is completed, secondary compression takes place under constant effective stress while there is no more excess pore water pressure to dissipate. During secondary compression, the coefficient of secondary compression C_α is defined as the change in void ratio per log cycle of time and is given by:

$$C_\alpha = \frac{\Delta e}{\Delta \log t} \qquad \text{(B18.4)}$$

where Δe and $\Delta \log t$ are the changes in void ratio and log time, respectively, during a time interval. The void ratios computed at the end of consolidation due to each pressure increment and the unloading steps can be plotted against the corresponding effective vertical stresses σ'_v (log scale) to generate the $e - \log \sigma'_v$ plot which is one of the main objectives of the consolidation test. From this plot, the *compression index* C_c, *recompression index* C_r (also known as *swelling index* C_s), and the *preconsolidation pressure* σ'_p can be obtained. The overconsolidation ratio *OCR* is defined as:

$$OCR = \frac{\sigma'_p}{\sigma'_{v0}} \qquad \text{(B18.5)}$$

where σ'_{v0} = in situ effective vertical stress on the sample that can be estimated from the sample depth, water table depth, and the unit weights of the overlying soils. For normally consolidated soils, *OCR* = 1. For overconsolidated soils, *OCR* is greater than 1. The compression index is often proportional to the liquid limit, natural water content, or initial void ratio. Terzaghi and Peck (1967) suggested that:

$$C_c = 0.009\,(LL - 10) \text{ for undisturbed clay} \qquad \text{(B18.6)}$$

and

$$C_c = 0.007\,(LL - 10) \text{ for remolded clays} \qquad \text{(B18.7)}$$

Floating ring (Figure B18.1b) and fixed ring (Figure 18.1c) are two types of oedometers used in consolidation tests, both consisting of the same components. In a floating ring, the specimen is compressed from top and bottom toward the center, and, hence, the specimen movement relative to the ring is also toward the center. In a fixed ring, the compression and the specimen movement within the ring are both toward the bottom. The fixed ring has the advantage of allowing independent measurements of permeability to be carried out by a falling head permeability test through the standpipe at any pressure increment. The side friction is reduced by 50% in a floating ring where the maximum friction occurs at mid-height.

Procedure:

The procedure outlined here applies to saturated or nearly saturated intact or remolded specimens.

1. *Soil specimen size*: The test can be carried out on an undisturbed soil specimen at least 50 mm in diameter, 12 mm in thickness, and a diameter-to-height ratio of at least 2.5.

Larger diameter-to-height ratios minimize the effects of side friction at the oedometer ring wall. Side friction can be reduced by applying a coat of silicone grease on the oedometer wall.

2. *Specimen preparation*: Place a good quality undisturbed specimen in the oedometer ring and determine the mass of the specimen m_1. This specimen may be obtained by trimming a tube sample or a block sample, or by pushing the oedometer ring into the clay. Measure the diameter (D) and thickness (H_0) of the specimen. Determine the water content (w_0) and specific gravity (G_s) from the trimmings. From these, the initial void ratio (e_0) can be determined.

3. *Porous stones*: Porous stones are made of noncorrosive material such as silicon carbide or aluminum oxide, for example. Use a filter paper (No. 54 Whatman) to separate the porous stones and the specimen, thus preventing clogging of the pores in the porous stone.

4. *Specimen assembly and seating load*: Place the specimen in the ring with a porous stone and filter paper at the top and bottom and apply a small seating pressure of about 5 kPa or less. This would ensure that the loading arrangement is securely in contact with the specimen and will prevent any swelling of the specimen when inundated with water. This should cause negligible consolidation of the specimen. In the case of very soft clays, the seating pressure can be reduced. Initialize the deformation indicator (e.g., dial gage or LVDT) and inundate the specimen in water, which will remain in water for the entire duration of the test, thus ensuring the specimen is saturated. It may be necessary to increase the seating load to prevent swelling.

5. *Loading*: Load increment ratio (*LIR*) is defined as the ratio of the applied pressure increment to the current pressure on the specimen. *LIR* of 1.0 is suggested as the norm. In other words, the pressure on the specimen would be doubled in the subsequent increment. ASTM D2435 suggests pressures of 12, 25, 50, 100, 200 kPa. This may be continued to 400, 800, and 1600 kPa before the specimen is unloaded. Lower *LIR* can be used in very compressible soils or in the vicinity of the preconsolidation pressure to better define it.

6. *During consolidation*: While consolidation takes place due to a specific pressure increment, record the change in the specimen thickness over time at approximately 0.1, 0.25, 0.5, 1, 2, 4, 8, 15, 30 minutes, and 1, 2, 4, 8, and 24 hours. It is common to see data acquisition systems that take millions of readings at specified time intervals.

7. *Unloading*: With each pressure increment maintained for 24 hours to ensure full consolidation, the subsequent increment is added. Once the maximum pressure required by the client is reached, unloading can take place in decrements by halving the pressures or, if necessary, reducing the successive pressures by one-fourth.

8. *Unload-reload cycle (optional)*: To define the unloading path and compute the recompression (swelling) index, it is often recommended to carry out an unload-reload cycle

after passing the preconsolidation pressure. This is required to draw the field *virgin consolidation line*, in the case of overconsolidated clays, as suggested by Schmertmann (1955).

9. At the end of the consolidation test, extrude the specimen from the oedometer, determine the water content (w_f), and the mass and thickness (H_f) of the specimen.

Notes:

1. *Porous stones*: The permeability of the porous stone must be at least one order of magnitude greater than that of the specimen. It is a common practice to boil the porous stone for 10 minutes or more to remove entrapped air bubbles. Between tests, the porous stones should be stored in de-aired water.

2. *Porous stone diameter*: In a floating ring oedometer, both porous stones should be 0.2 to 0.5 mm smaller in diameter than the inner diameter of the ring. In a fixed ring oedometer, only the upper porous stone has to be smaller. It is preferable that the stones are tapered slightly with the larger diameter in contact with the soil.

3. *Porous stone material*: Sintered bronze, carborundum, porous brass, and porous stainless steel are common materials used for making porous stones.

4. ASTM D2435 suggests that the porous stones and the specimen should be assembled such that there is no drastic change in the water content of the specimen. For example, use dry porous stones for dry and expansive soils, damp porous stones for partially saturated specimens, and saturated porous stones for saturated specimens that have a low affinity to water.

5. *Pore water pressure measurements*: Some oedometers drain from one end and have the other end sealed to facilitate pore water pressure measurement during consolidation. Measuring pore water pressure is an effective way of defining the *end of primary consolidation*. These oedometers may also have the facility to apply back pressure during consolidation, which is an effective way to ensure saturation throughout consolidation.

6. *Saturation*: There is a better chance that the specimen would be saturated at the end of the test than at the beginning. Therefore, it is better to use the final measurements (w_f, H_f, etc.) and back calculate the initial void ratio e_0 that is required in the computations. Further, the initial water content is taken from the trimmings whereas the final water content is taken directly from the test specimen, which is more reliable.

Datasheet:

A simple datasheet for recording all necessary data for developing $e - \log \sigma'_v$ plot is given in Table B18.1. This table includes: (a) the project details; (b) the specimen height, water content, void ratio, dry density, and degree of saturation at the beginning and end of the consolidation test; and (c) $e - \sigma'_v$ values for each pressure increment.

Table B18.1 Consolidation test data to generate e – log σ'_v plot

Sample description:	High plastic clay with significant sand
Sample location:	Kirwan Hospital
Sample no.:	BH4-4.5m
Date:	12-March-2005
Tested by:	Warren O'Donnell
Notes:	

In situ and associated specimen data:

Specimen depth from GL (m)　　　4.40　　　Water table depth (m)　　　1.80

Estimated in situ effective overburden stress on the specimen (kPa)　　　45.0

Soil classification:　　CH–High plastic clay with some sand

Undrained shear strength (kPa):　　33.0 (from UU Triaxial)

In situ water content (%)　60.5

LL	55	PL	26	PI	29	G_s	2.69

Specimen measurements (Initial and final):

Specimen diameter (mm)	48.1	x-section area (cm^2)	18.171
Initial thickness (mm)	20.0	Final thickness (mm)	16.429
Initial water content (%)	63.6	Final water content (%)	47.9
Initial wet mass (g)	59.08	Final wet mass (g)	53.34
Dry mass of solids (g)	36.11	Dry mass of solids (g)	36.06
Dry density (g/cm^3)	0.994	Dry density (g/cm^3)	1.208
Void ratio (e_0)	1.707	Void ratio (e_f)	1.227
Degree of saturation (%)	100.2	Degree of saturation (%)	105.0

Void ratio—effective stress plot:

	Dial gage readings (mm)							
σ'_v (kPa)	d_0	d_{100}	d_f	H_o (mm)	ΔH (mm)	H_f (mm)	Δe	e
2.0			0.000	20.000				1.707
20.0	NA	NA	0.084	20.000	0.084	19.916	0.0114	1.696
37.9	0.091	0.132	0.178	19.916	0.094	19.822	0.0127	1.683
69.7	0.175	0.340	0.441	19.822	0.263	19.559	0.0356	1.647
137.4	1.470	1.475	1.797	19.559	1.356	18.203	0.1835	1.464
272.6	1.800	3.280	3.588	18.203	1.791	16.412	0.2424	1.221
543.0	3.610	5.030	5.212	16.412	1.624	14.788	0.2198	1.002
137.4	NA	NA	4.729	14.788	−0.483	15.271	−0.065	1.067
20.0	NA	NA	3.571	15.271	−1.158	16.429	−0.157	1.224

The following information can be included in the test report:

- Soil profile at the site, water table location, and specimen depth (to estimate the initial effective overburden stress σ'_{v0})
- Atterberg limits, specific gravity, and natural water content
- Description and classification of the soil
- Initial and final values of water content, degree of saturation, and dry density of the specimen
- Void ratio (e) versus the logarithm of effective vertical stress (σ'_v) plot (see Figure B18.2) and values of the compression index (C_c), the recompression index (C_r), and preconsolidation pressure (σ'_p)
- Plots of deformation against logarithm of time (Casagrande's method) or square root of time (Taylor's method) for all pressure increments
- Coefficient of consolidation (c_v), coefficient of volume compressibility (m_v), and coefficient of secondary compression (C_α) plotted against effective vertical stress. ASTM D2435 suggests an alternate plot of logarithm of c_v against the void ratio.

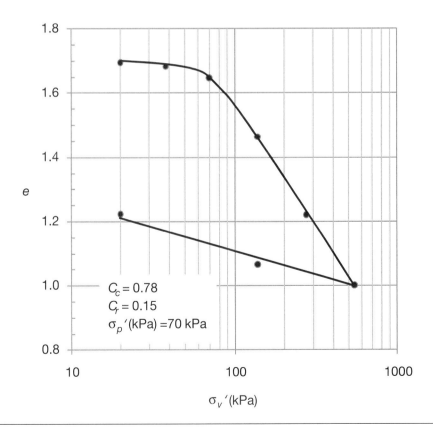

$C_c = 0.78$
$C_r = 0.15$
$\sigma_p'(kPa) = 70\ kPa$

Figure B18.2 e versus σ'_v plot

Analysis:

Initial void ratio: Determination of the initial void ratio (e_0) is the starting point for preparing the data for $e - \log \sigma'_v$ plot. This is the void ratio of the specimen just before applying the first increment that can be determined from the water content of the trimmings (w_0), initial height (H_0), diameter (D), mass (m_1) of the specimen, and the specific gravity of the soil grains (G_s). The initial dry density of the specimen can be computed as:

$$\rho_d = \frac{m_1}{\pi D^2 H_0} \times \frac{1}{(1 + w_0)} \tag{B18.8}$$

The initial void ratio can be computed from the dry density using the following equation:

$$e_0 = \frac{G_s \rho_w}{\rho_d} - 1 \tag{B18.9}$$

The water content used in these computations is from the trimmings and not from the specimen. Hence, there can be a slight difference. A better way of computing e_0 is to back calculate from the *final* measurements where the final water content and the mass of dry solids can be accurately determined at the end of the test.

$e - \sigma'_v$ *data*: For every pressure increment, the change in the dial gage reading gives the change in the specimen thickness. The corresponding change in void ratio can be determined by:

$$\frac{\Delta e}{1 + e_i} = \frac{\Delta H}{H_i} \tag{B18.10}$$

where ΔH = change in thickness, Δe = change in void ratio, e_i = void ratio just before applying the pressure increment, and H_i = the specimen thickness just before the pressure increment. Using Equation B18.10, the changes in void ratio and, hence, the values of void ratios at the end of consolidation due to each pressure increment can be determined. This can continue for the unloading too. *In the calculations shown in Table B18.1, no attempt is made to separate the secondary compression and consolidation in computing Δe.*

Coefficient of volume compressibility, m_v: For each pressure increment, the coefficient of volume compressibility (m_v) can be computed from Equation B18.3. It is a measure of compressibility of the clay with values typically ranging from 0.05 to 1.5 MPa^{-1} for most clays. In the normally consolidated pressure range (i.e., $\sigma'_v > \sigma'_p$), m_v and C_c are related by:

$$m_v = \frac{0.434 \, C_c}{(1 + e_0) \sigma'_{\text{average}}} \tag{B18.11}$$

In the overconsolidated range, C_c in this equation is replaced by C_r. m_v is, in fact, the reciprocal of the *constrained modulus*, also known as the *oedometer modulus*.

Preconsolidation pressure, σ'_p: From the $e - \log \sigma'_v$ plot, the preconsolidation pressure should be determined by Casagrande's (1936) graphical procedure. The compression index measured from the laboratory $e - \log \sigma'_v$ plot can be less than the actual in situ value. Schmertmann (1955) suggested a graphical procedure to determine the *in situ* virgin consolidation line, the slope of which gives the true compression index.

Coefficient of consolidation, c_v: Corresponding to the data given in Table B18.1, the deformation (i.e., dial gage reading) and time readings for the pressure increment from 272.6 kPa to 543.0 kPa are shown in Table B18.2. To determine c_v, ASTM D2435 recommends two of the standard procedures that are also described in most geotechnical engineering textbooks. They are: (1) Casagrande's logarithm of time method and (2) Taylor's square root of time method. These methods can also be used to indirectly define the end of the consolidation process.

(1) Casagrande's logarithm of time method

Here, the dial gage reading (or specimen thickness) is plotted against the logarithm of time as shown in Figure B18.3.

Table B18.2 Time—Dial gage reading for σ'_v from 272.6 kPa to 543.0 kPa

Time	Dial gage reading (mm)
0⁻	3.590 (just before loading increment)
1.5 s	3.676
15 s	3.690
30 s	3.718
1 min	3.756
2 min	3.806
4 min	3.884
8 min	3.983
16 min	4.130
32 min	4.330
60 min	4.562
141 min	4.853
296 min	5.027
429 min	5.086
459 min	5.095
680 min	5.141
1445 min	5.204
1583 min	5.212

Note: Dial gage reading at the beginning of the consolidation test was 0.000 mm

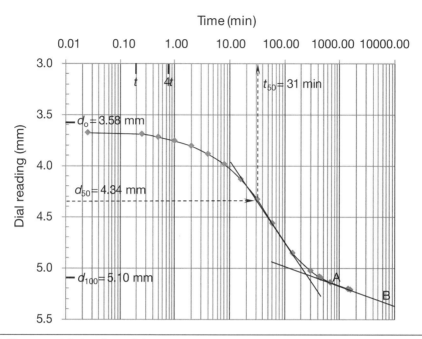

Figure B18.3 Casagrande's log time plot

First, it is required to identify the two linear segments of the plot in the middle and at the end. The intersection of these two straight lines defines the dial gage reading corresponding to 100% consolidation, d_{100}. Now, it is necessary to find the dial gage reading corresponding to the beginning of the consolidation process d_0, which could not have been identified precisely. This can be accomplished by selecting two times (t and $4t$) within the initial parabolic part of the plot and noting the difference in dial gage readings between the two times. Subtracting this difference from the dial gage reading corresponding to time t defines d_0. The dial gage reading corresponding to 50% consolidation d_{50} is obtained by averaging d_0 and d_{100}. The time at 50% consolidation t_{50} can be read off the plot (see Figure B18.3 for the graphical procedures). The time factor T_{50} at average consolidation of 50% is 0.197. Using the appropriate value for the maximum length of drainage path H_{dr} at t_{50}, the coefficient of consolidation can be determined from:

$$c_v = \frac{T_{50} H_{dr}^2}{t_{50}}$$

(B18.12)

For specimens drained from top and bottom, H_{dr} is half the specimen thickness; for specimens drained from one side only, H_{dr} is the same as the specimen thickness. For the data presented in Tables B18.1 and B18.2 and Figure B18.3:

$$c_v = \frac{0.197\left(\dfrac{20.0 - 4.34}{2}\right)^2}{31 \times 60} = 0.0065 \text{ mm}^2/s$$

Here, as in most consolidation tests, the specimen is doubly drained.

(2) Taylor's square root of time method

In Taylor's square root of time method, the dial gage reading is plotted against square root of time, as shown in Figure B18.4. The dial gage reading corresponding to the beginning of the consolidation process d_0 is obtained by extending the straight line segment of the plot until it meets the vertical $t = 0$ axis.

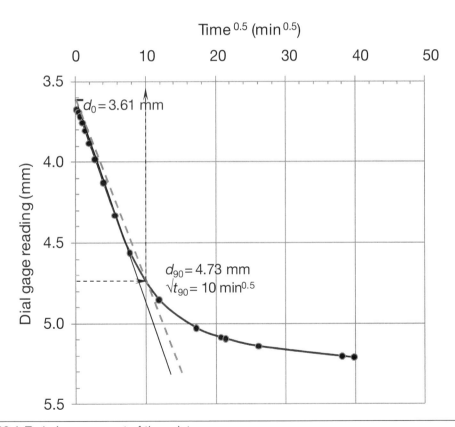

Figure B18.4 Taylor's square root of time plot

A second straight line (dashed in Figure B18.4) is drawn through d_0 such that the abscissa is 1.15 times larger than the previous line. The intersection of the second line with the deformation—square root of time plot—defines 90% consolidation from which $\sqrt{t_{90}}$ can be determined. The time factor T_{90} for 90% average degree of consolidation is 0.848:

$$c_v = \frac{T_{90}H_{dr}^2}{t_{90}} \quad\quad (B18.13)$$

The average value of the maximum length of drainage path H_{dr} is still obtained at 50% consolidation. The points on the curve corresponding to 50 or 100% consolidation can be identified from the fact that the deformation axis is linear. In Figure B18.4, d_{50} can be computed as:

$$d_{50} = 3.61 + \frac{5}{9} \times (4.73 - 3.61) = 4.232 \text{ mm}$$

And, hence, $H_{dr} = (20 - 4.232)/2 = 7.884$ mm. Substituting $H_{dr} = 7.884$ mm, $t_{90} = 100$ minutes in Equation B18.13, c_v can be determined as 0.0088 mm^2/s.

Coefficient of secondary compression, C_α: The coefficient of secondary compression (C_α) can be determined from the tail end of the dial gage reading versus log time plot (Casagrande) shown in Figure B18.3, using Equation B18.4. Here, the changes Δe and $\Delta \log t$ will be calculated between the two points A (700 min, 5.15 mm) and B (8000 min, 5.31 mm). The void ratio at A and B can be calculated as:

$$e_A = e_0 - \frac{\Delta H}{H_0}(1 + e_0) = 1.707 - \frac{5.15}{20.0}(1 + 1.707) = 1.010$$

$$e_B = e_0 - \frac{\Delta H}{H_0}(1 + e_0) = 1.707 - \frac{5.31}{20.0}(1 + 1.707) = 1.988$$

There, $C_\alpha = \dfrac{1.010 - 0.988}{\log\left(\frac{8000}{700}\right)} = 0.021$.

Permeability, k: Permeability is better measured through a falling head permeability test under the applied pressures, using a fixed ring type oedometer. An indirect estimate is possible through the following equation:

$$k = c_v m_v \gamma_w \quad\quad (B18.14)$$

This estimate is not reliable as it depends on the validity of the one-dimensional consolidation theory and the assumptions. For the data given in Tables B18.1 and B18.2, when the pressure is increased from 272.6 kPa to 543 kPa, the permeability can be estimated as:

$$m_v = \frac{\Delta H}{H \times \Delta \sigma} = \frac{1.624}{16.412 \times (543.0 - 272.6)} = 0.366 \times 10^{-3} \text{ kPa}^{-1} = 0.366 \text{ MPa}^{-1}$$

$$\therefore k = \left(0.65 \times 10^{-8}\, \frac{m^2}{s}\right)\left(0.366 \times 10^{-6}\, \frac{m^2}{N}\right)\left(9810\, \frac{N}{m^3}\right) = 2.33 \times 10^{-11}\, \text{m/s}$$

Cost: US$600–US$700 (Extra US$200 for unload-reload cycle)

B19 Direct Shear Test

Objective: To determine the drained shear strength parameters of a soil through a direct shear test

Standards: ASTM D3080
AS 1289.6.2.2
BS1377-7

Introduction

Soils are generally modeled by the Mohr-Coulomb constitutive model at failure where the shear strength τ_f is related to the normal stress σ on the failure plane by:

$$\tau_f = c + \sigma \tan \phi \qquad (B19.1)$$

where c = cohesion and ϕ= friction angle of the soil. The equation defines the failure envelope in τ-σ plane. The shear strength parameters, c and ϕ, are different for drained and undrained loading situations. There is no provision to control the drainage during shearing in a standard direct shear apparatus. Therefore, it is generally recommended that a direct shear test be carried out under drained conditions where the loading rate is slow enough to ensure that there is no buildup of excess pore water pressures. Nevertheless, it is possible to shear the specimen under undrained conditions through a faster loading rate.

An oversimplified schematic diagram of a direct shear test setup is shown in Figure B19.1 where the soil specimen is contained within a small, rigid metal box having a square or circular cross section in plan view. The box is split horizontally with a tiny gap between the upper and lower sections. The soil specimen contained within the box is typically 60 to 100 mm in width and 20 to 25 mm in thickness. The bottom half is fixed to the base, whereas the top half can be moved relative to the bottom one (or the opposite, commonly), thus shearing the soil specimen along the horizontal *failure plane* shown by the dashed line in Figure B19.1.

A normal load N is applied vertically which results in a normal stress σ applied within the specimen and on the failure plane. The normal stress σ is given by N/A where A is the cross sectional area of the specimen. A shear load S is applied on the upper box, and increased gradually from zero. The shear stress τ on the failure plane is given by S/A. Two displacement dial gages are attached to the apparatus to measure the vertical and horizontal displacements (δ_v and δ_h) of the upper half of the specimen. By plotting the shear stress τ against the horizontal displacement δ_h, the *shear stress at failure* (i.e., *shear strength*) τ_f can be identified. This procedure can be repeated for three different values of normal stresses. It must be remembered that failure can be defined in terms of peak or residual shear stresses. *Peak shear stress* is the maximum value of the shear stress during the test; *residual shear stress* is the shear stress at very large strain that can be

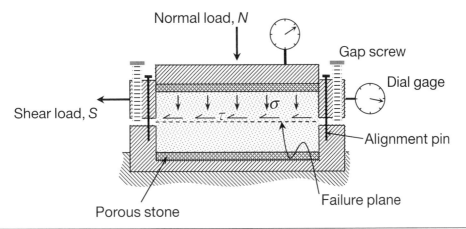

Figure B19.1 Schematic diagram of direct shear test

taken as the value at the end of the test. For practical purposes, peak shear stress can be taken as the *ultimate shear stress*. The corresponding values of τ_f can be plotted against σ, thus defining the failure envelope (Equation B19.1) from which c and ϕ can be determined.

Procedure:

1. *Soil specimen*: In the case of clayey soils, the test can be carried out on intact, remolded, or compacted specimens. Sand specimens are generally reconstituted to specific relative densities.
2. *Shear box*: The shear box is generally made of stainless steel, bronze, or aluminum with provisions for drainage from the top and bottom. The box is split along a horizontal plane into an upper and lower half, separated by a very thin gap, and fitted together with two vertical alignment pins. One of the halves is fixed and the other is moveable. Two or three gap screws are used to control the gap between the two halves.
3. *Porous stones*: Porous stones are provided at the top and bottom of the specimen to facilitate drainage.
4. *Displacement dial gages*: Displacement dial gages or transducers can be used for measuring horizontal and vertical displacements. Horizontal ones may have a stroke length of 15 mm and a precision of ± 0.025 mm. Vertical ones will have a better precision of ± 0.0025 mm.
5. *Shear load measuring device*: A proving ring or load cell accurate to 2.5 N, or 1% of the shear load at failure, is recommended for measuring the shear load S.
6. Assemble the shear box, keeping the two alignment pins in place to ensure that the top and bottom halves are aligned properly. Ensure that the gap screws are wound back so that they don't protrude, and that the gap between the two halves is minimal. Place

the specimen of known mass (m_1) and water content (w_1) sandwiched between the two porous stones and the loading platen at the top. Determine the specimen dimensions. Attach the vertical displacement dial gage.

7. Apply the desired normal load N and allow complete consolidation before applying the shear load. This can be ensured by recording the specimen compression against the time and following Casagrande's or Taylor's procedures discussed under the consolidation test.

8. Remove the vertical alignment pins and use the gap screws to open the gap between the two halves to 0.64 mm.

9. *Shearing*: Shearing should be carried out slowly such that there is no buildup of excess pore water pressure. The duration of shearing (i.e., time to failure) t_f can be estimated as 50 times t_{50} where t_{50} is the time taken for 50% consolidation. For free draining clean sands, $t_f = 10$ minutes, and for dense sands with more than 5% fines, $t_f = 60$ minutes. Shearing rate can be estimated on the basis that the horizontal displacement δ_h at failure is in the order of 5 to 12 mm, the higher values applicable for loose sands and normally consolidated clays. Attach the horizontal displacement dial gage and set the loading rate using the gear wheels or motor speed.

10. It is common to use a data acquisition system for a direct shear test. During shearing, the following readings are taken: shear load S, shear displacement δ_h, and vertical displacement δ_v. Continue shearing well beyond failure.

11. Repeat the above steps for three or more normal loads.

Notes:

1. For fine grained soils, the container holding the shear box can be filled with water for the entire duration of the test.

2. The direct shear test is generally strain-controlled, rather than stress-controlled. The advantage of a strain-controlled test is that the peak and residual states can be defined more precisely in the case of a strain-softening soil.

Datasheet:

A simple datasheet for a direct shear test on sands is given in Table B19.1 (data after Lambe 1951). The following information can be included in the test report:

- Soil classification, location, and project name
- Type of shear device used
- Initial and final water contents
- Initial and final dry densities
- Loading (strain) rate and shearing time for each load
- Plots of τ versus δ_h and δ_v versus δ_h for each specimen (better on the same plot)

Table B19.1 Datasheet for direct shear test on sand (after Lambe 1951)

Soil description	Brown color fine uniform quartz sand, subrounded grains
Sample no.	CA-1-30 (Lambe 1951)
Location	Union Falls, Maine, USA
Date	18-Sep-50
Tested by	WCS

Specimen mass, m_1 (g)	112.8	Initial water content, w_1 (%)		0
Spec. gravity of the grains	2.67			
Specimen dimensions (mm):				
Length	76.2	Width	76.2	Thickness 12.7
Area (cm²)	58.06			Volume (cm³) 73.74
Normal load (N)	3337,40	Normal stress σ(kPa)	574.8	

Time	δ_h (mm)	δ_v (mm)	S (N)	τ (kPa)
0_s	0.00	0.000	0.0	0.0
15_s	0.05	−0.015	422.7	72.8
30_s	0.23	−0.051	952.3	164.0
45_s	0.48	−0.081	1357.2	233.7
1 min	0.74	−0.086	1590.8	274.0
1.25 min	1.02	−0.117	1813.3	312.3
1.75 min	1.85	−0.117	2176.0	374.7
2 min	2.29	−0.107	2280.5	392.8
	3.05	−0.066	2385.1	410.8
	3.56	−0.053	2411.8	415.4
	4.06	−0.041	2431.8	418.8
3 min	4.57	−0.030	2447.4	421.5
	5.08	−0.020	2447.4	421.5
	5.59	−0.015	2442.9	420.7
	6.10	−0.013	2425.1	417.7
4 min	6.60	−0.013	2425.1	417.7
	7.11	−0.013	2422.9	417.3
	7.62	−0.013	2407.4	414.6
	8.13	−0.013	2398.5	413.1
5 min	8.64	−0.013	2382.9	410.4
	9.14	−0.015	2385.1	410.8
	9.65	−0.018	2391.8	411.9
	10.16	−0.023	2402.9	413.8
6 min	10.67	−0.025	2402.9	413.8
	11.18	−0.030	2407.4	414.6
6.5 min	11.68	−0.030	2402.9	413.8

- Ultimate and peak (if different) shear stresses for each normal stress and the two failure envelopes.

Analysis:

The initial dry density can be computed as:

$$\rho_d = \frac{m_1}{1 + w_1} \tag{B19.2}$$

where the initial water content is expressed as a decimal number (e.g., 0.155 instead of 15.5%). The final dry density can be computed in a similar manner from the final water content and the specimen dimensions.

The normal stress during shearing was computed as 574.8 kPa. As the specimen is sheared, with increasing horizontal displacement δ_h, there is a gradual reduction in the shearing area A' which may be computed as $A' = L(L-\delta_h)$. AS 1289.6.2.2 recommends using this reduced area in computing the shear and normal stresses. This correction is not recommended in ASTM D3080 and was not carried out for the data presented in Table B19.1 and Figure B19.2. The plots of τ and δ_v against δ_h are shown in Figure B19.2. The peak and ultimate shear strengths are identified as 425 kPa and 420 kPa, which are not very different. In case of dense sands and overconsolidated clays, there can be a significant drop in shear strength from the peak to ultimate states. Only in normally consolidated clays and loose sands, they are about the same where the peak occurs at the ultimate state.

By plotting the shear strength against the normal stress, from the data for three or more normal loads, the failure envelopes for peak and residual states can be defined, and, from these, the values of c and ϕ can be determined. In terms of effective stresses, ϕ_{peak} is greater than $\phi_{ultimate}$, and $c_{ult} \approx 0$. In granular soils, ϕ_{peak} is significantly influenced by the relative density; $\phi_{ultimate}$ is independent of the relative density. Typical values of ϕ_{peak} and $\phi_{ultimate}$ as reported in the literature are summarized in Table B19.2. McCarthy (2007) suggests slightly higher values than those reported in Table B19.2. Quoting 30° as the typical value for $\phi_{ultimate}$, Lambe (1951) gave significantly higher values for ϕ_{peak}: 37° to 60° for dense, well graded coarse sand and 33° to 45° for dense, uniform fine sand.

McCarthy (2007) suggests that the friction angle determined from a direct shear test is 1° to 4° greater than that from a triaxial test. The friction angle increases with angularity due to

Table B19.2 Typical values of ϕ_{peak} and $\phi_{ultimate}$ (after Lambe and Whitman 1979)

Soil	ϕ_{peak} (°)	$\phi_{ultimate}$ (°)
Medium dense to dense, nonplastic silt	28-34	26-30
Medium dense to dense fine-medium uniform sand	30-36	26-30
Medium dense to dense well graded sand	34-46	30-34
Medium dense to dense sand and gravel	36-48	32-36

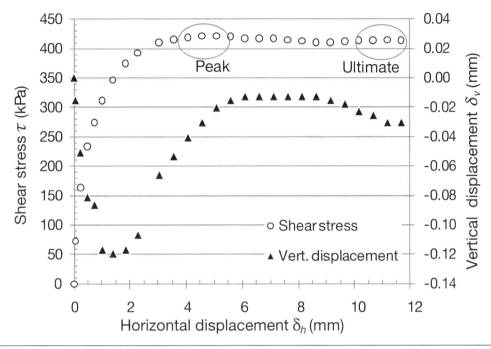

Figure B19.2 Plots of τ and δ_v against δ_h for data in Table B19.1

better interlocking between the grains. From rounded to angular grains, the friction angle can increase by 5°. Well graded granular soils have a higher friction angle than poorly graded ones due to denser packing between the grains.

Cost: US$600–US$700

B20   Consolidated Undrained Triaxial Test

Objective: To determine the shear strength parameters (c and ϕ) of a soil from a triaxial test

Standards: ASTM D4767
AS 1289.6.4.2

Introduction

Failure in soils occurs generally in shear mode along a slip surface. The shear stress and normal stress at a point on the failure surface are related by:

$$\tau_f = c + \sigma \tan \phi \qquad \text{(B20.1)}$$

where τ_f = shear strength (shear stress at failure), c = cohesion, σ = normal stress, and ϕ = friction angle. It can be seen from Equation B20.1 that the shear strength consists of two separate components that are derived from cohesion and friction. While the cohesive component is independent of the normal stress, the frictional component is proportional to the normal stress.

The triaxial test is carried out on a cylindrical soil specimen with a length (L) to diameter (D) ratio of 2.0 to 2.5 with a minimum diameter of 33 mm. The test consists of two stages: (1)

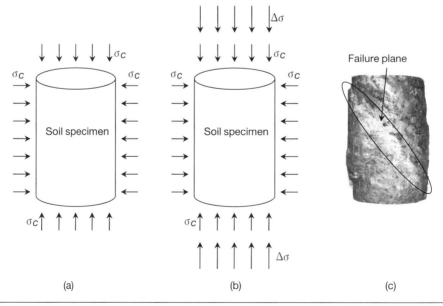

Figure B20.1 Two stages of loading: (a) isotropic confinement; (b) additional vertical stress; and (c) photograph of a failed specimen (Photograph: N. Sivakugan)

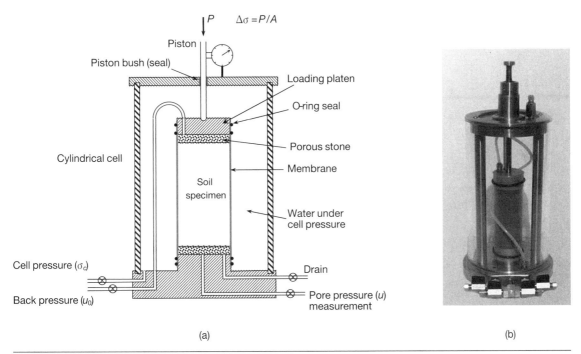

Figure B20.2 Triaxial test setup: (a) schematic and (b) photograph (Photograph: N. Sivakugan)

application of isotropic confining cell pressure right around the specimen and (2) application of an additional vertical stress and increasing it till failure (see Figure B20.1). Figure B20.2 shows a schematic diagram and a photograph of a triaxial cell assembly. The vertical wall of the soil specimen is enclosed in an impermeable rubber membrane with o-rings providing a firm seal at the top and bottom. The porous stones at the top and bottom facilitate the drainage from the specimen.

While applying the isotropic cell pressure (Figure B20.1a), the specimen can be allowed to consolidate by opening the drainage valve. Keeping the valve closed ensures that the specimen does not consolidate during the application of the cell pressure. The specimen is said to be *consolidated* or *unconsolidated*, depending on whether the drainage valve is open or closed during this stage.

In the second stage (Figure B20.1b), the additional vertical stress (often incorrectly called *deviator stress*) $\Delta\sigma$ is applied through a piston load P that is increased to failure. At any time during loading, $\Delta\sigma = P/A$, where A is the cross-sectional area of the specimen. Here again, depending on whether the drainage valve is open or closed, the specimen is said to undergo *drained loading* or *undrained loading*. In drained loading, the rate of applying $\Delta\sigma$ must be slow enough to ensure that there is no buildup of excess pore water pressure within the specimen. The rate of loading is rather quick for undrained loading where pore water pressure develops during the loading.

When the specimen fails, the minor and major principal stresses in terms of total and effective stresses are given by:

$$\sigma_{3f} = \sigma_c \qquad (B20.2)$$

$$\sigma_{1f} = \sigma_c + \Delta\sigma_f \qquad (B20.3)$$

$$\sigma'_{3f} = \sigma_c - u_f \qquad (B20.4)$$

$$\sigma'_{1f} = \sigma_c + \Delta\sigma_f - u_f \qquad (B20.5)$$

where $\Delta\sigma_f$ is the value of $\Delta\sigma$ at failure and u_f is the pore water pressure at failure. By testing three or more specimens subjected to different confining cell pressures, Mohr circles at failure can be drawn and the failure envelopes can be drawn from which c and ϕ can be determined. The failure plane within the specimen is inclined at an angle of $45 + \phi/2$ to horizontal (Figure B20.1c)

Catering for different needs, the following three types of triaxial tests are commonly carried out in geotechnical laboratories:

1. Consolidated drained (CD) triaxial test
2. Consolidated undrained (CU) triaxial test [ASTM D4767, AS 1289.6.4.1]
3. Unconsolidated undrained (UU) triaxial test [ASTM D2850, AS 1289.6.4.1]

A CD triaxial test is carried out with the drainage valve open during both stages. The data is often analyzed in terms of effective stresses and c' and ϕ' can be determined. The isotropic consolidation in the first stage and the very slow loading in the second stage require a few days to test a specimen, which makes a CD test expensive and rather time consuming. A CU triaxial test is similar to a CD test in the first stage where the specimen is consolidated. In the second stage, however, the drainage valve is closed and the additional vertical stress is applied rather quickly, which induces pore pressure within the specimen. With pore water pressure measurements, the analysis can be carried out in terms of effective stresses to determine c' and ϕ'. A CU test is significantly faster than a CD test and is often the preferred method for determining c' and ϕ'. CD and CU tests can be carried out on granular and cohesive soils. For normally consolidated clays and granular soils $c' = 0$. Even for overconsolidated clays, c' is relatively small, often less than 10 kPa. In other words, in terms of effective stresses, the shear strength is mainly frictional.

The UU triaxial test is carried out with the drainage valve closed during both stages. In Stage 1, no consolidation takes place during the application of cell pressure, thus maintaining the initial water content of the specimen. Pore water pressure develops during both stages, but is not measured and, hence, the effective stresses are unknown. The analysis is generally carried out in terms of total stresses, where the failure envelope is horizontal for saturated clays (i.e., $\phi = 0$). The cohesion measured from the envelope is known as the *undrained shear strength* c_u.

In CD and CU triaxial tests, a *back pressure* (u_0) is often applied to the pore water to dissolve any air bubbles within the system, thus ensuring full saturation. It can be applied during consolidation under the cell pressure and also during the drained loading. Here, the consolidation and/or the drained loading can take place against the constant back pressure u_0.

Procedure:

1. *Soil specimen:* The test can be carried out on remolded and compacted specimens or intact specimens collected from the site. Determine the length (L), diameter (D), mass (m_1), and initial water content (w_0) of the specimen. See the relevant standards for details on specimen preparation and mounting it on the pedestal within the triaxial cell.
2. Fill the triaxial cell with de-aired water without leaving any space within the cell.
3. *Saturation:* The first attempt to saturate the specimen can be through the application of vacuum and flushing de-aired water through the system. Especially in the case of low permeability soils, it is often necessary to apply a back pressure to drive any remaining air into the water. Back pressures required to attain a certain degree of saturation are shown in Figure B20.3.
4. Bring the axial load piston into contact with the loading platen at the top of the specimen without applying any significant load (e.g., less than 0.5% of the failure load).
5. *Backpressure:* Backpressure should not be applied in one step; it has to be applied in increments along with the cell pressure. Allow sufficient time between increments for the equalization of pore water pressure throughout the specimen. During back pressure saturation, it is necessary to ensure that the effective cell pressure (the difference

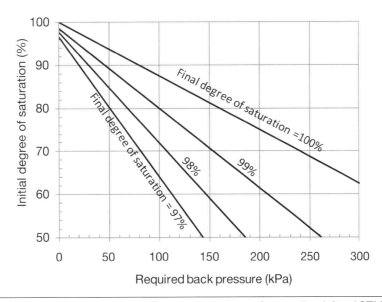

Figure B20.3 Back pressures required to attain various degrees of saturation (after ASTM D4767)

between cell and back pressures) is positive but very small (e.g., less than 35 kPa) so that there will be negligible consolidation of the specimen.

6. *B-check:* Measuring Skempton's *B*-parameter is a good way to ensure saturation. *B*-parameter is defined as:

$$B = \frac{\Delta u}{\Delta \sigma_c} \tag{B20.6}$$

where Δu is the increase in pore water pressure corresponding to the increase of $\Delta \sigma_c$ in the confining pressure. In an ideal situation, *B* is expected to be close to unity for a saturated soil. A value greater than 0.95 is acceptable. It is a common practice to carry out a *B*-check at the end of saturation and report the value.

7. *Consolidation:* Apply the required effective consolidation pressure with the appropriate back pressure and let the specimen consolidate. The water squeezed out of the specimen is collected in a volume change burette connected to the drainage line.

8. *Axial loading:* The typical loading rate is ½ to 1% per minute in strain-controlled tests or about ⅛ of the compressive strength per minute in stress-controlled tests. Record the axial deformation of the specimen, the vertical load *P*, and pore pressure at 0.1, 0.2 to 1.0%, and then sufficient readings to define the stress-strain curve. The loading may be continued to one of the following three: (1) 15% axial strain, (2) drop in $\Delta \sigma$ by 20%, or (3) post-peak axial strain of additional 5%.

9. At the end of the test, remove the axial load and reduce the cell and back pressures to zero. Remove the rubber membrane and the porous stones and determine the water content of the specimen. Take a photograph and prepare a sketch of the failed specimen, showing the failure plane.

Notes:

1. The theoretical expression for Skempton's *B*-parameter is:

$$B = \frac{1}{1 + \dfrac{nC_v}{C_{sk}}} \tag{B20.7}$$

where n = porosity, C_{sk} = compressibility of soil skeleton (kPa^{-1}), and C_v = compressibility of the pore fluid, air, and water when partially saturated (kPa^{-1}). Compressibility of water is 5.4×10^{-3} kPa. C_{sk} is generally significantly greater than C_v and, hence, *B* is about 1, especially for soft to medium-stiff soils. In the case of very stiff soils, where C_{sk} is not very high, *B* can be less than unity even when fully saturated.

2. It is common to consolidate the specimen to an effective consolidation pressure σ_c' that is equal to the in situ overburden pressure, thus replicating the field situation.

3. In drained loading in a CD test, the strain rate of 0.06% per hour was used on an undisturbed Boston blue clay specimen that had permeability of 2×10^{-7} cm/s (Lambe 1951).

4. Undrained shear strength increases with the strain rate.

Datasheet:

A simple datasheet for a consolidated undrained triaxial test on a silty clay soil is given in Table B20.1. The specimen was consolidated under a cell pressure of 378.5 kPa with no back pressure. It can be seen that the undrained shearing of the specimen took 26 minutes and 8 seconds when

Table B20.1 Datasheet for consolidated undrained triaxial test (after Lambe 1951)

Time (min-s)	σ_c (kPa)	ΔL (mm)	ε (%)	A (cm^2)	P (N)	$\Delta\sigma$ (kPa)	u (kPa)	σ_3' (kPa)	σ_1' (kPa)
0-0	378.5	0.000	0.00	37.84	0.0	0	1.4	377.1	377.1
	378.5	0.076	0.05	37.86	461.3	121.8	21.4	357.1	479.0
	378.5	0.152	0.09	37.88	603.6	159.4	46.9	331.6	491.0
	378.5	0.559	0.34	37.97	780.6	205.6	64.1	314.4	520.0
	378.5	0.787	0.48	38.03	825.5	217.1	86.9	291.6	508.7
3-0	378.5	1.803	1.11	38.27	931.9	243.5	147.6	230.9	474.5
	378.5	2.337	1.44	38.39	951.0	247.7	179.3	199.2	446.9
	378.5	3.124	1.92	38.58	964.8	250	202.7	175.8	425.8
	378.5	3.505	2.16	38.68	969.7	250.7	211.0	167.5	418.2
	378.5	3.886	2.39	38.77	974.6	251.4	217.2	161.3	412.7
	378.5	4.420	2.72	38.9	974.6	250.5	224.1	154.4	404.9
	378.5	4.928	3.03	39.03	973.2	249.4	227.5	151.0	400.3
	378.5	5.461	3.36	39.16	970.6	247.9	240.6	137.9	385.7
6-20	378.5	5.994	3.69	39.29	968.3	246.5	244.1	134.4	380.9
	378.5	6.528	4.02	39.43	961.2	243.8	251.0	127.5	371.3
	378.5	7.061	4.34	39.56	957.7	242.1	254.4	124.1	366.1
8-20	378.5	7.595	4.67	39.7	954.5	240.5	256.5	122.0	362.5
	378.5	8.128	5.00	39.83	950.1	238.5	264.1	114.4	352.9
9-25	378.5	8.661	5.33	39.97	943.0	235.9	259.9	118.6	354.5
	378.5	9.169	5.64	40.1	938.5	234	264.1	114.4	348.4
10-25	378.5	9.703	5.97	40.24	924.3	229.7	266.8	111.7	341.3
	378.5	10.236	6.30	40.39	921.2	228.1	272.4	106.1	334.2
11-29	378.5	10.770	6.63	40.53	915.4	225.9	275.1	103.4	329.3
	378.5	11.278	6.94	40.66	912.3	224.4	278.6	99.9	324.3
	378.5	11.811	7.27	40.81	907.8	222.5	279.9	98.6	321.0
	378.5	12.344	7.59	40.95	905.2	221	281.3	97.2	318.2
13-35	378.5	12.878	7.92	41.1	897.6	218.4	281.3	97.2	315.6
	378.5	13.411	8.25	41.24	894.0	216.8	281.3	97.2	313.9
	378.5	13.945	8.58	41.39	891.8	215.5	278.6	99.9	315.4
	378.5	14.478	8.91	41.54	888.7	213.9	287.5	91.0	304.9
	378.5	15.011	9.23	41.69	884.3	212.1	288.9	89.6	301.7
	378.5	16.078	9.89	42	877.6	209	291.0	87.5	296.5
17-45	378.5	17.145	10.55	42.3	864.2	204.3	291.0	87.5	291.8
	378.5	18.186	11.19	42.61	845.1	198.3	291.0	87.5	285.9
19-51	378.5	19.253	11.84	42.93	809.5	188.6	297.2	81.3	269.9
	378.5	20.295	12.48	43.24	768.2	177.7	306.8	71.7	249.3
	378.5	21.387	13.16	43.58	712.6	163.5	320.6	57.9	221.4
	378.5	22.479	13.83	43.91	705.5	160.6	317.9	60.6	221.3
26-08	378.5	25.654	15.78	44.93	691.7	153.9	311.0	67.5	221.5

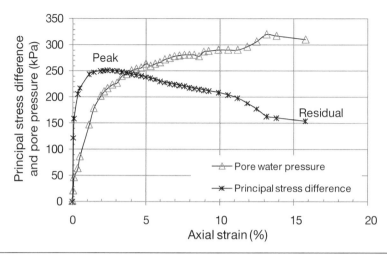

Figure B20.4 Plot of principal stress difference and pore water pressure against axial strain

the axial strain was 15.1%. The plot of principal stress difference ($\Delta\sigma$) and pore water pressure against the axial strain is shown in Figure B20.4.

The following information can be included in the test report:

- Visual description of the soil and whether the specimen is intact, compacted, or other
- LL, PL, G_s, and grain size data
- Initial dimensions and water content of the specimen
- Back pressure applied
- Method of saturation and B-parameter at the end of saturation
- Strain rate during undrained loading
- Axial strain at failure
- Plots of principal stress difference and pore water pressure against axial strain
- Sketch or photograph of the specimen at failure
- Mohr circles at failure in terms of effective and total stresses, along with the corresponding envelopes

Analysis:

Change of dimensions during consolidation and undrained loading:

During consolidation, the specimen length and volume decrease by ΔL_c and ΔV_c, respectively. The new cross-sectional area A_c at the end of consolidation is given by:

$$A_c = \frac{V_0 - \Delta V_c}{L_0 - \Delta L_c} \tag{B20.8}$$

where L_0 = initial length of the specimen and V_0 = initial volume of the specimen.

The average cross-sectional area of the specimen would increase during undrained loading. Assuming that the specimen is saturated and hence there is no volume change:

$$A_c L_c = A (L_c - \Delta L)$$

where L_c = length of the specimen at the end of consolidation and at the start of undrained loading, A_c = cross-sectional area of the specimen at the start of undrained loading, ΔL = change in specimen length at a specific time during loading, and A = current cross-sectional area at that time. Therefore:

$$A = \frac{A_c}{\left(1 - \dfrac{\Delta L}{L_0}\right)} = \frac{A_c}{1 - \varepsilon} \tag{B20.9}$$

where ε is the axial strain of the specimen. Equation B20.9 should be used to determine the new cross-sectional area at any instant during the undrained loading.

Skempton's pore water pressure parameters A and B:

During any undrained loading, Skempton (1954) showed that the change in pore water pressure is given by:

$$\Delta u = B[\Delta \sigma_3 + A (\Delta \sigma_3 - \Delta \sigma_3)] \tag{B20.10}$$

where A and B are Skempton's pore pressure parameters and $\Delta \sigma_1$ and $\Delta \sigma_3$ are the largest and smallest change in the principal stresses. During an increase in the cell pressure of $\Delta \sigma_c$ (i.e., $\Delta \sigma_1 = \Delta \sigma_3 = \Delta \sigma_c$), the change in pore pressure Δu is given by $B \Delta \sigma_c$. Therefore, B can be computed from Equation B20.6.

Failure and Mohr circles:

Failure can be defined in terms of peak or residual (or ultimate) values (see Figure B20.4). The values of principal stresses at failure, in terms of effective stresses and total stresses, can be computed and corresponding Mohr circles can be drawn. Failure envelopes can be drawn tangent to the Mohr circles, from which c', ϕ', c, and ϕ can be determined.

Cost: US$1000–US$1200

B21 Unconsolidated Undrained Triaxial Test

Objective: To determine the undrained shear strength parameters (c_u and ϕ_u) and stress-strain relationship of a clay using a triaxial test setup

Standards: ASTM D2850
AS 1289.6.4.1

Introduction

When analyzing the *short-term* stability of a clay soil, it is often assumed that the clay is undrained, and the analysis is carried out in terms of total stresses where the shear strength parameters are the undrained shear strength c_u and the undrained friction angle that is zero. The unconsolidated undrained triaxial test is one of the better ways of determining c_u.

The UU triaxial test is the simplest form of all three triaxial tests discussed in the last test. Here, the drainage valve remains closed during the application of the cell pressure *and* while the additional vertical load is applied to shear the specimen. Therefore, the specimen remains at the same water content throughout the test. If the specimen is saturated, the specimen volume remains the same throughout the test in spite of any change in shape.

The UU triaxial test is analyzed in terms of total stresses and, hence, the pore water pressures are never measured. When the specimens are at the same water content and are saturated, irrespective of the confining pressure, the specimens will fail under the same vertical load. Therefore, the Mohr circles in terms of total stresses are of the same size, and the failure envelope is horizontal as shown in Figure B21.1a. Here, the friction angle ϕ_u in terms of total stresses is zero, which is a standard attribute of saturated clays under undrained conditions. However, when the specimens are partially saturated, they do consolidate under the confining pressures and the strength increases with the confining pressures and the failure envelope is no longer horizontal, as shown in Figure B21.1b, and can be curved as well.

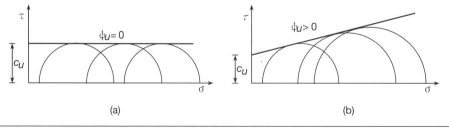

Figure B21.1 Failure envelopes in undrained loading: (a) saturated specimens and (b) partially saturated specimens

Procedure:

1. *Soil specimen:* The test can be carried out on remolded and compacted specimens or intact specimens collected from the site. The cylindrical specimen should have a minimum diameter of 33 mm and a length to a diameter ratio of 2.0 to 2.5. Determine the initial length (L_0), initial diameter (D_0), mass (m_1), and initial water content (w_0) of the specimen. See the relevant standards for details on specimen preparation and mounting it on the pedestal within the triaxial cell.
2. Fill the triaxial cell with de-aired water without leaving any space within the cell.
3. Bring the axial load piston into contact with the loading platen at the top of the specimen without applying any significant load (e.g., less than 0.5% of the failure load).
4. Apply the cell pressure and allow 10 minutes for the specimen to equilibrate before applying the additional axial load.
5. *Axial loading:* The loading is generally strain-controlled in a UU triaxial test. The typical loading rate is 0.3 to 1% per minute; the slower rates for brittle materials and faster ones for the ductile materials. The maximum principal stress difference (i.e., peak vertical load) is reached in 15 to 20 minutes. Record the axial deformation of the specimen and the vertical load P at 0.1, 0.2 to 0.5%, and then at every 0.5 to 3% axial strain, and at every 1% thereafter. The loading may be continued to one of the following three: (1) 15% axial strain, (2) drop in $\Delta\sigma$ by 20%, or (3) post-peak axial strain of an additional 5%.
6. At the end of the test, remove the axial load and reduce the cell pressure to zero. Remove the rubber membrane and the porous stones and determine the water content of the specimen. Take a photograph and prepare a sketch of the failed specimen, showing the failure mode (e.g., shear plane, bulging, etc.).

Datasheet:

A simple datasheet for an unconsolidated undrained triaxial test on a firm clay soil is given in Table B21.1. The cell pressure for the test was 100 kPa. The principal stress difference peaked at 68 kPa, under an axial strain of 4.2% (see Figure B21.2).

The following information can be included in the test report:

- Visual description of the soil and whether the specimen is intact, compacted, or other
- *LL*, *PL*, G_s, and grain size data
- Initial dimensions and water content of the specimen
- Strain rate during undrained loading
- Axial strain at failure
- Stress-strain plot
- Sketch or photograph of the specimen at failure
- Mohr circles at failure in terms of total stresses and the failure envelope

Table B21.1 Datasheet for unconsolidated undrained triaxial test

Initial dimensions:

Specimen length, L_0 (mm)	165.1
Specimen diameter, D_0 (mm)	71.0
x-sectional area, A_0 (cm^2)	39.59
Volume, V_0 (cm^3)	653.7

σ_c (kPa)	ΔL (mm)	ε (%)	P (N)	A (cm^2)	$\Delta\sigma$ (kPa)
100	0	0.000	0	39.59	0.00
100	0.15	0.091	32.5	39.63	8.20
100	0.3	0.182	64.4	39.66	16.24
100	0.45	0.273	95	39.70	23.93
100	0.6	0.363	128.9	39.74	32.44
100	0.75	0.454	155	39.77	38.97
100	0.9	0.545	170.2	39.81	42.75
100	1.5	0.909	196.3	39.95	49.13
100	2.5	1.514	223.4	40.20	55.57
100	3.5	2.120	241.9	40.45	59.80
100	4.5	2.726	255.6	40.70	62.80
100	5.5	3.331	269.4	40.96	65.78
100	7.25	4.391	282.5	41.41	68.22
100	8.5	5.148	281.1	41.74	67.34
100	10	6.057	270	42.14	64.07
100	12	7.268	260.1	42.70	60.92
100	14	8.480	253.4	43.26	58.58
100	16	9.691	251.6	43.84	57.39
100	18	10.902	249.5	44.44	56.15
100	20	12.114	246.3	45.05	54.66
100	22	13.325	241.5	45.68	52.87

Analysis:

During the undrained loading, the cross-sectional area of the specimen increases and can be calculated from Equation B21.1.

$$A = \frac{A_0}{\left(1 - \dfrac{\Delta L}{L_0}\right)} = \frac{A_0}{1 - \varepsilon} \tag{B21.1}$$

Here, ε = axial strain of the specimen, A_0 = average initial cross section of the specimen, ΔL = change in length from the beginning of axial loading, and L_0 = initial length of the specimen.

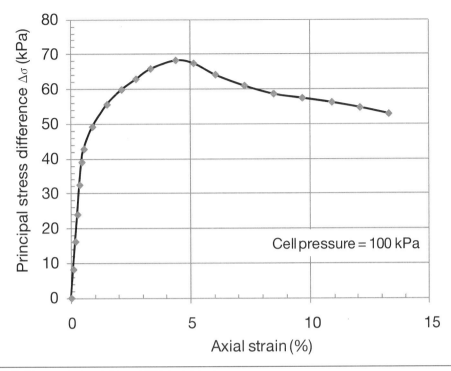

Figure B21.2 Stress-strain plot in a UU triaxial test

Failure and Mohr circle:

At failure, the cell pressure in the minor principal stress, and hence $\sigma_{3f} = \sigma_c$. If the additional vertical stress applied at failure is $\Delta\sigma_f$, the major principal stress σ_{1f} is given by $\sigma_c + \Delta\sigma_f$. Using these two values of the Mohr circle at failure can be drawn in terms of total stresses. The test can be repeated at three or more different confining pressures, and a Mohr circle can be drawn for each confining pressure. The failure envelope is drawn tangent to the Mohr circles from which c_u and ϕ_u are determined.

Cost: US$200–US$300

B22   Unconfined Compression Test

Objective: To determine the unconfined compressive strength of a cohesive soil

Standards: ASTM D2166

Introduction

Unconfined compression or an uniaxial compression test is a special case of a triaxial test. It is carried out without any confining pressure and the triaxial cell assembly is not required, making the test procedure simpler. It is carried out on saturated clay specimens that can stand unsupported. The clay specimen is loaded axially where the load is increased to failure. The loading is carried out rather quickly, leaving little time for drainage. Therefore, it is reasonable to assume that the test is carried out under undrained conditions. During the loading, the axial shortening of the specimen and the applied load are measured at certain intervals from which the stress-strain plot can be generated and the unconfined compressive strength, also known as uniaxial compressive strength, is determined. From a good quality stress-strain plot, the undrained Young's modulus of the clay can also be determined. This is a simpler but less reliable method than the UU triaxial test to determine the undrained shear strength of clay.

The unconfined compressive strength of a clay can be estimated by pocket penetrometers or a pocket vane shown in Figure B22.1 These devices are quite popular in the field as they enable quick estimates of the undrained shear strength of clay, costing literally nothing. However, these estimates should be used with caution; they are unreliable. In the pocket penetrometers (Figure B22.1a), the penetrating piston diameter can be varied to suit the consistency of the clay. Similarly, in the pocket vane (Figure 22.1b) the vane size can be changed also, and in some devices the spring can be changed too.

Procedure:

1. *Soil specimen:* The test can be carried out on remolded and compacted specimens or intact specimens collected from the site. The cylindrical specimen should have a minimum diameter of 33 mm and a length to diameter ratio of 2.0 to 2.5. Determine the length (L), diameter (D), mass (m_1), and initial water content (w_0) of the specimen. See the relevant standards for details on specimen preparation.
2. Place the specimen between the two loading platens in the loading frame such that it is centered on the bottom platen. Lower the upper platen until it is in contact with the specimen without applying any load. Initialize the displacement measurement device.
3. The loading is generally strain-controlled. Apply the vertical load at a strain rate of ½ to 2%/min. Continue the loading until the applied loads start decreasing or at least

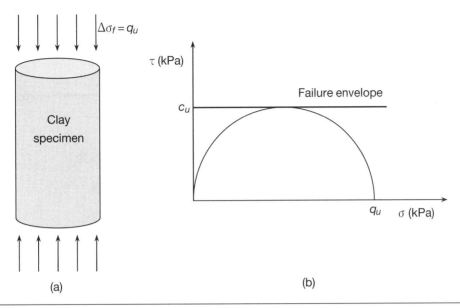

Figure B22.1 Unconfined compression test: (a) loading and (b) Mohr circle

15% strain is reached. Select the strain rate such that the time to failure is less than 15 minutes. During the loading, record the applied load and the axial shortening of the specimen at appropriate intervals so that there are at least 15 points to define the stress-strain curve.

4. At the end of the test, remove the axial load and remove the specimen. Take a photograph and prepare a sketch of the failed specimen, showing the failure mode (e.g., shear plane, bulging, etc.).

5. Determine the water content using the entire specimen.

Datasheet:

A datasheet similar to that used in a UU triaxial test can be used for recording load-deformation data. The following information can be included in the test report:

- Visual description of the soil and whether the specimen is intact, compacted, or other
- Atterberg limits and grain size data
- Initial dry density and water content
- Computed value of degree of saturation (requires G_s)
- Initial dimensions
- Strain rate during undrained loading
- Axial strain at failure
- Stress-strain plot
- Sketch or photograph of the specimen at failure

Analysis:

A schematic diagram of a clay specimen subjected to unconfined compression test and the Mohr circle at failure are shown in Figure 22.2a and b, respectively. At failure, the major and minor principal stresses are given by: $\sigma_3 = 0$ and $\sigma_1 = q_u$ (see Figure 22.2a), where q_u is the additional vertical stress applied at failure and is known as the unconfined compressive strength.

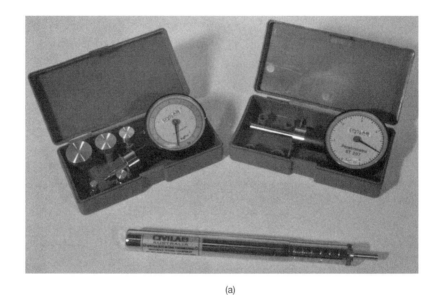

(a)

(b)

Figure B22.2 Devices for q_u estimates: (a) pocket penetrometers and (b) pocket vane (Photograph: N. Sivakugan)

Since the test is carried out under undrained conditions, if the clay is saturated, the failure envelope will be horizontal (i.e., $\phi_u = 0$) as shown in Figure 22.2b.

It can be seen that the undrained shear strength c_u and unconfined compressive strength q_u of a saturated clay are related by:

$$c_u = \frac{1}{2}\, q_u \tag{B22.1}$$

In unsaturated specimens where $\phi_u > 0$, it can be shown from the Mohr circle that:

$$c_u = \frac{1 - \sin \phi_u}{2 \cos \phi_u} q_u \tag{B22.2}$$

In plotting the stress-strain curve, the cross-section area has to be corrected as in the CU and UU tests, using Equation B21.1.

Cost: US$100–US$125

This book has free material available for download from the
Web Added Value™ resource center at *www.jrosspub.com*

Rock Testing

Part C

INTRODUCTION

The International Society of Rock Mechanics (ISRM) was founded in 1962 at Karlsruhe University by Professor Leopold Mueller. It appointed the *Commission on Standardization of Laboratory and Field Tests on Rock* in 1967, which later became *The Commission on Testing Methods*. The commission proposed *suggested methods* for various rock tests that have been adopted worldwide and were published from time to time in the *International Journal of Rock Mechanics and Mining Sciences & Geomechanics Abstracts*, Pergamon Press, Oxford, England. These were compiled by Professor Ted Brown (1981) of the University of Queensland, Australia, as the ISRM *yellow book*. This was later updated by Professor Ulusay of Hacettepe University, Turkey, and Professor Hudson, formerly of Imperial College, United Kingdom in 2007 as a *blue book* that is a one-stop shop for all relevant ISRM suggested methods for rock testing.

The test procedures for rock described in this section are based mainly on the ISRM suggested methods with references to ASTM International (ASTM) and Australian Standards (AS) as appropriate.

Drill cores are generally brought to the laboratory in a sample tray or core box as shown in Figure C.1. The cores are labeled with a borehole number and depth and stacked in an orderly manner in the sample tray. The cores are recovered from boreholes, using a core barrel that has a diamond bit attached at the tip. The common core sizes and the designation are given in Table C.1.

The core barrel may consist of single, double, or triple tubes to minimize the disturbance. Total core recovery (R) is the length of the recovery expressed as the percentage of the drill depth. Rock quality designation (RQD) is a modified measure of core recovery in a borehole that is defined as (Deere 1964):

$$RQD\,(\%) = \frac{\Sigma \ lengths \ of \ core \ pieces \ longer \ than \ 100 \ mm}{Total \ length \ of \ the \ core \ run} \times 100 \qquad (C.1)$$

The RQD is a simple and inexpensive way to recognize low quality rock zones that may require further investigation. The RQD, corresponding descriptions of in situ rock quality, and the allowable foundation pressures as given by Peck et al. (1974) are summarized in Table C.2. ISRM recommends RQD be computed for double tube NX cores of 54 mm diameter.

Figure C.1 Drill cores from the site in a sample tray (Photograph: N. Sivakugan)

Table C.1 Core size designations and nominal diameters

Symbol	Nominal core diameter (mm)	(inches)
AQ	27	1–1/16
BQ	36.5	1–7/16
NQ	47.6	1–7/8
HQ	63.5	2–1/2
PQ	85.0	3–11/32
EX	22.2	7/8
AX	30.2	1–3/16
BX	41.3	1–5/8
NX	54.0	2–1/8

Table C.2 RQD, in situ rock quality description and allowable bearing pressure (after Peck et al. 1974)

RQD (%)	Rock quality	Allowable bearing pressure (MPa)
0–25	Very poor	1-3
25–50	Poor	3-6.5
50–75	Fair	6.5-12
75–90	Good	12-20
90–100	Excellent	20-30

The behavior of a *rock mass* is governed by the presence of discontinuities (e.g., joints), their orientations, strength, and so forth. *Intact rock* is the material between the discontinuities that is often tested in the laboratory. It requires good judgment to arrive at the bigger picture using the lab data and the discontinuities present in the field. We generally test the intact rock in the laboratory and then extrapolate to the rock mass in the field situation, considering the presence of discontinuities. Rock specimens are prepared as per the required standards (see Figure C.2).

Figure C.2 Rock specimen preparation—polishing the ends (Photograph: N. Sivakugan)

The *uniaxial compression test*, also known as the *unconfined compression test*, is the most common rock test for assessing the strength of intact rock and rock masses. In the literature, uniaxial compressive strength or unconfined compressive strength is denoted by σ_c, q_u, or *UCS*. It is the most commonly used parameter in rock characterization and designs and numerical modeling of rock mechanics problems. A general rule of thumb is that when σ_c exceeds 1 MPa, it is considered rock; below this threshold value the material may be treated as a soil.

C1 Water Content

Objective: To determine the water (moisture) content of a rock sample consisting of a collection of rock fragments

Standards: ASTM D2216
ISRM (1979b)*

Introduction

Water content influences rock behavior. It is a common practice to specify the water content of the rock specimen when it is tested. The test is similar to the water content determination of soils.

Procedure:

1. Determine the mass m_1 of a clean and dry container.
2. Select a representative sample containing at least 10 lumps of rock each having a mass of at least 50 g and a minimum dimension of 10 times the maximum grain size.
3. Place the sample in the container and determine the mass m_2.
4. Place the sample and the container in an oven at 105 to 110°C and dry to constant mass m_3. (Note: The sample can be allowed to cool in a dessicator for 30 minutes before determining the mass.)

Datasheet:

A simple datasheet for this test, prepared in Excel, is shown in Table C1.1. The water content should be reported to the nearest 0.1%.

The other data recorded may include:

- Lithologic (macroscopic) description of the rock
- Sample source

Table C1.1 Water content measurements

Sample no.	m_1 (g)	m_2 (g)	m_3 (g)	w (%)	Comments
JR012	56.05	643.2	622.6	3.6	24 hrs in oven
JR019	56.05	678.4	958.5	3.3	24 hrs in oven
JR032	56.05	623.9	603.7	3.7	28 hrs in oven

- Whether in situ water content and, if so, precautions taken to preserve the water content (preferably within 1% of the in situ value) during sampling and storage
- Standards followed and deviations, if any

Note:

ASTM standard is the same for both soils and rocks.

Analysis:

The *water content of the rock sample* is defined as:

$$w\ (\%) = \frac{m_2 - m_3}{m_3 - m_1} \times 100 \qquad (C1.1)$$

Cost: US$5 to US$10

C2   Density and Porosity

Objective: To determine the density and porosity of

a. a regular rock specimen in the form of a right cylinder
b. a sample consisting of a collection of irregular rock fragments

Standards: ISRM (1979b)*
AS 4133.2.1.1 & AS 4133.2.1.2

Introduction

The bulk density and porosity of a rock are related by the phase relations that are also used for soils. In the case of a regular rock specimen, the volume V_t is determined from the dimensions that are measured precisely. For irregular rock fragments, the volume is determined using Archimedes' buoyancy principle that states that the up thrust on a submerged body is equal to the weight of the liquid displaced.

The porosity n is the ratio of void volume V_v to the total volume V_t expressed as a percentage. The void volume is computed from phase relations using the dry (m_s) and saturated (m_t) masses of the rock sample.

Procedure:

a. Regular rock specimen

1. Select a regular specimen with a mass of at least 50 g and a minimum dimension of at least 10 times the largest grain size.
2. Measure the dimensions to 0.1 mm using a caliper. Compute the volume V_t.
3. Immerse the specimen in a container of water and apply a vacuum of 800 Pa or more for at least 1 hour with periodic agitation to remove the trapped air. Record the water temperature to the nearest 1°C.
4. Remove the specimen from the water and dry the surface using a moist cloth.
5. Determine the mass of the saturated surface-dry specimen m_t.
6. Dry the specimen to constant mass (e.g., until the reduction is less than 0.1% of the initial mass) at 105 to 110°C, allowing it to cool in a dessicator for 30 minutes before determining the mass m_s.
7. Compute the pore volume, porosity, and dry density using Equations C2.1 through C2.3.
8. Repeat for three or more specimens.

b. Irregular rock fragments

1. Select a representative sample comprised of 10 or more rock lumps, each having a mass of at least 50 g and a minimum dimension greater than 10 times the largest grain size. Wash the sample to remove any dust.
2. Saturate the sample by immersing in water under a vacuum of at least 800 Pa for at least an hour with periodic agitation to remove any trapped air. Record the water temperature to the nearest 1°C.
3. Submerge the sample and determine the saturated-submerged mass m' to 0.1 g. This can be done as follows:
   ○ Determine the mass (m_{bs}) of a wire basket or a perforated container submerged in the immersion bath.
   ○ Transfer the sample to the wire basket in the immersion bath and determine the saturated-submerged mass (m_{ss}) of the sample plus basket.
   ○ Calculate the saturated-submerged mass m' as $m' = m_{ss} - m_{bs}$.
4. Remove the sample from the water and dry the surface using a moist cloth.
5. Determine the mass of the saturated surface-dry sample m_t to 0.1 g.
6. Dry the specimen to constant mass (e.g., until the reduction is less than 0.1% of the initial mass) at 105°C, allowing it to cool in a dessicator for 30 minutes before determining the mass m_s.
7. Compute the total volume V_t using Equation C2.4.
8. Compute the pore volume, porosity, and dry density using Equations C2.1 through C2.3.

Datasheet:

Datasheets for this test, prepared in Excel, are shown in Tables C2.1 and C2.2 for regular and irregular specimens. For regular specimens, individual test results for three or more specimens should be reported along with the average results. Density should be reported to the nearest 10 kg/m^3 or 0.01 t/m^3 and porosity to the nearest 0.1%.

The other data recorded may include:

- Lithologic (macroscopic) description of the rock
- Sample source
- Individual determinations and the average values of both density and porosity
- Specific gravity of the grains
- Standard followed and deviations, if any

Table C2.1 Density and porosity measurements for regular cylindrical specimens

Specimen no.	Length L (mm)	Diameter D (mm)	Volume V_t (cm³)	Dry mass m_s (g)	Sat. mass m_t (g)	Void vol. V_v (cm³)	Porosity n (%)	Dry unit wt γ_d (kN/m³)
BH5-22m1	157.2	63.1	491.6	1435.4	1506	70.2	14.3	28.6
BH5-22m2	158.3	63.1	495.0	1442.3	1519	76.6	15.5	28.6
BH5-22m3	160.1	63.1	500.7	1465.1	1535	70.1	14.0	28.7
Average							14.6	28.6

Table C2.2 Density and porosity measurements for irregular rock fragments

Specimen no.	Sample mass			Total vol. V_t (cm³)	Void vol. V_v (cm³)	Porosity n (%)	Dry unit wt γ_d (kN/m³)
	Submerged m' (g)	Saturated m_t (g)	Dry m_s (g)				
S21	374.8	561.2	543.2	186.4	18.0	9.7	28.6
S26	384.2	579.3	563.7	195.1	15.6	8.0	28.3
Average						8.8	28.5

Analysis:

The pore volume V_v is given by:

$$V_v = \frac{m_t - m_s}{\rho_w} \qquad (C2.1)$$

The porosity of the rock n is defined as:

$$n\,(\%) = \frac{V_v}{V_t} \times 100 \qquad (C2.2)$$

The dry density ρ_d of the rock is defined as:

$$\rho_d = \frac{m_s}{V_t} \qquad (C2.3)$$

From Archimedes' principle, the total volume of the sample consisting of irregular rock fragments can be determined from the saturated and submerged masses as:

$$V_t = \frac{m_t - m'}{\rho_w} \qquad (C2.4)$$

Cost: US$50 (regular specimen) and $75 (irregular rock fragments)

C3 Uniaxial Compressive Strength

Objective: To determine the uniaxial compressive strength σ_c, Young's modulus E and Poisson's ratio υ of a rock sample

Standards: ASTM D7012
ISRM (1979a)*
AS 4133.4.2.1

Introduction

Clays under undrained conditions are generally analyzed using the total stress parameters c_u and ϕ_u. Here, c_u is the undrained shear strength and ϕ_u is the friction angle in terms of total stresses. In saturated clays $\phi_u = 0$. The unconfined compression test (Test B22) is one of the many ways of deriving the undrained shear strength of a clay. The unconfined compressive strength (UCS) of a clay, denoted often by q_u in geotechnical literature, is twice the undrained shear strength c_u when $\phi_u = 0$.

The same principle holds in rocks too. Uniaxial compressive strength, often denoted by σ_c in rock mechanics literature, is the most used rock strength parameter in rock mass classification and rock engineering designs. Unlike undrained clays, the friction angle of a rock specimen is not zero, and, hence, the Mohr-Coulomb failure envelope is not horizontal. It can be shown from a Mohr circle that:

$$\sigma_c = \frac{2\,c\cos\phi}{1 - \sin\phi} \tag{C3.1}$$

where c and ϕ are the cohesion and friction angle of the rock.

The test is quite simple and the interpretation is fairly straightforward. A cylindrical rock core, with a length/diameter ratio of 2.5 to 3.0, is subjected to an axial load that is increased to failure. Uniaxial compressive strength is the maximum load carried by the specimen divided by the cross-sectional area. By generating the stress-strain plot during the loading, Young's modulus can be computed. By measuring diametric or circumferential strains during loading, Poisson's ratio can be measured. Hawkes and Mellor (1970) discussed various aspects of the UCS laboratory test procedure in great detail.

Procedure:

1. The test specimen diameter should be at least of NX core size (54 mm), and the length should be approximately equal to 2.5 to 3.0 (ASTM suggests 2.0 to 2.5) times the

diameter. The test specimens should be cut and prepared using clean water. The ends of the test specimens should be flat with a maximum variation of 0.02 mm and should be parallel to each other and at right angles to the longitudinal axis. The sides of the specimens should be smooth and free of abrupt irregularities and straight to within 0.3 mm over the full length of the specimen. Use of capping material or end surface treatment is not permitted.

2. Take two measurements of the diameter at right angles at the top, middle, and bottom of the core to the nearest 0.1 mm. The average of the six readings should be used as the original diameter d_0 of the specimen.

3. Determine the original height of the specimen l_0 to the nearest 1.0 mm.

4. If Young's modulus is required, provisions should be made to record the vertical deformation (or strain) of the specimen. If Poisson's ratio is required, provisions should be made to record the change in the diameter of the specimen. At least 10 evenly spaced readings are required during the loading to define the axial and diametric stress-strain curves shown in Figure C3.1. Currently, with data acquisition systems readily available, it is common to have continuous measurements during loading. The displacement or strain measurement devices include electrical resistance strain gages, linear variable differential transformers, compressometers, optical devices, and other suitable measuring devices that are robust and stable with strain sensitivity of the order of 5×10^{-6}. Their design should be such that the average of two axial and two circumferential strain measurements, equally spaced, can be determined for every load increment.

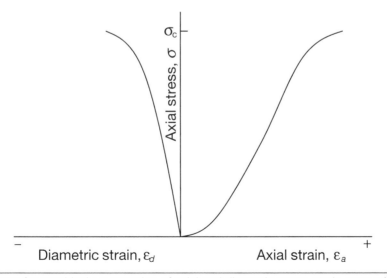

Figure C3.1 Format for graphical presentation of axial and diametric stress-strain curves (adapted from ISRM 1978b)

5. Load the specimen at a constant rate so that it fails in 5 to 15 minutes. Alternatively the stress rate should be in the range of 0.5 to 1.0 MPa/s. To prevent injury from flying rock fragments on failure, a protective shield should be placed around the test specimen (see Figure C3.2).

6. Record the maximum load to within 1%.

7. If the load and deformation are continuously recorded, the stress-strain plots can be generated. The tangent (E_t) or secant (E_s) Young's modulus can be determined from the

Figure C3.2 Photograph of a UCS test setup with protective shield (Photograph: N. Sivakugan)

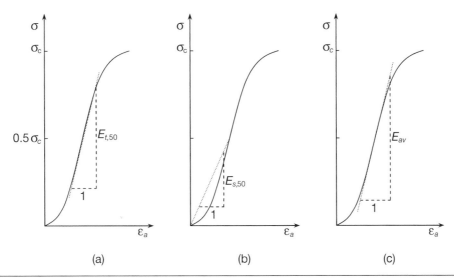

Figure C3.3 Methods for calculating Young's modulus E from axial stress-strain curve: (a) tangent modulus; (b) secant modulus; and (c) average modulus (adapted from ISRM 1978B)

stress-strain plot at a certain percentage (e.g., 50%) of the peak strength, as shown in Figures C3.3a and C3.3b. It is common to measure tangent and secant Young's modulus at 50% of σ_c.

Alternatively, an average Young's modulus E_{av} can be determined as the slope of the straight line portion of the stress-strain plot (Figure C3.3c). These are expressed in GPa.

Datasheet:

A simple datasheet for this test, prepared in Excel, is shown in Table C3.1.

Table C3.1 Uniaxial compression test data

Sample no.	d (mm)	l (mm)	Mass (g)	Volume (cm³)	Unit wt (kN/m³)	P (kN)	σ_c (MPa)
BH7_20.8m	53.8	140	908.3	318.71	28.0	320	140.9
BH7_21.4m	53.8	142	921.2	323.03	28.0	311	136.6
BH7_22.0m	53.9	142	922.3	322.87	28.0	333	145.8
BH7_22.5m	53.9	144	937.1	329.03	27.9	340	149.2
BH7_23.0m	53.8	140	905.9	317.12	28.0	322	141.5
Average σ_c (MPa)							142.8

The other data recorded may include:

- Lithologic (macroscopic) description of the rock
- Orientation of the axis of loading with respect to specimen anisotropy (e.g., bedding planes)
- Sample source, including location, depth, coring technique, storage history, and dates
- Number of specimens tested (at least five preferred)
- Specimen diameter and height
- Water content at the time of the test
- Test duration and loading rate
- Test date and testing machine details
- Modes of failure and strain at peak load
- Uniaxial compressive strength of each specimen expressed to three significant figures along with the average value
- If applicable, Young's modulus and Poisson's ratio for each specimen expressed to three significant figures along with the average values
- Standard followed and deviations, if any

Analysis:

There is little analysis involved in this test where the interpretation is rather straightforward. The axial strain ε_a is calculated from:

$$\varepsilon_a = \frac{\Delta l}{l_0} \tag{C3.2}$$

where Δl = the change in measured axial length defined positive when the length decreases; and l_0 = original axial length.

Diametric strain ε_d, which is the same as circumferential strain ε_c, is defined as:

$$\varepsilon_d = \frac{\Delta d}{d_0} = \varepsilon_c = \frac{\Delta C}{C_0} \tag{C3.3}$$

where Δd = change in diameter defined negative when diameter increases; d_0 = original diameter; ΔC = change in circumference; and C_0 = original circumference. UCS is useful in strength classification and in the characterization of the intact rock. Table C3.2 summarizes the rock classification system suggested by ISRM (1978b) with examples given by Hoek and Brown (1997).

Poisson's ratio V is a positive number in the range of 0.0 to 0.5 defined as the ratio of lateral strain to longitudinal strain. It may be calculated as:

$$v = -\frac{\text{Slope of axial stress-strain curve}}{\text{Slope of diametric stress-strain curve}} \tag{C3.4}$$

Table C3.2 Classification of soil and rock strengths (after ISRM 1978b; Hoek and Brown 1997)

Grade	Description	Field identification	σ_c or q_u (MPa)	Rock types
S1	Very soft clay	Easily penetrated several inches by fist.	< 0.025	
S2	Soft clay	Easily penetrated several inches by thumb.	0.025–0.05	
S3	Firm clay	Can be penetrated several inches by thumb with moderate effort.	0.05–0.10	
S4	Stiff clay	Readily indented by thumb, but penetrated only with great effort.	0.1–0.25[#]	
S5	Very stiff clay	Readily indented by thumbnail.	0.25[#]–0.50[#]	
S6	Hard clay	Indented with difficulty by thumbnail.	> 0.5[#]	
R0	Extremely weak rock	Indented by thumbnail.	0.25–1.0	Stiff fault gouge
R1	Weak rock	Crumbles under firm blows with point of geological hammer; can be peeled by pocket knife.	1–5	Highly weathered or altered rock
R2	Weak rock	Can be peeled by a pocket knife with difficulty; shallow indentations made by firm blow with a point of geological hammer.	5–25	Chalk, rock salt, and potash
R3	Medium strong rock	Cannot be scraped or peeled with a pocket knife; specimen can be fractured with a single firm blow of a geological hammer.	25–50	Claystone, coal, concrete, schist, shale, and siltstone
R4	Strong rock	Specimen requires more than one blow by geological hammer to fracture it.	50–100	Limestone, marble, phyllite, sandstone, schist, and shale
R5	Very strong rock	Specimen requires many blows of geological hammer to fracture it.	100–250	Amphibolite, sandstone, basalt, gabbro, gneiss, granodiorite, limestone, marble, rhyolite, and tuff
R6	Extremely strong rock	Specimen can only be chipped by a geological hammer.	> 250	Fresh basalt, chert, diabase, gneiss, granite, and quartzite

[#]Slightly different to the classification in geotechnical context

The volumetric strain ε_{vol} can be computed as:

$$\varepsilon_{vol} = \varepsilon_a + 2\varepsilon_d \qquad (C3.5)$$

Typical values of Poisson's ratios for different rock types are given in Table C3.3 (Gercek 2007).

The Young's modulus and Poisson's ratio thus measured are the two crucial parameters in defining rock behavior when it is assumed to behave as an elastic material. They are related to the bulk modulus K and shear modulus G by:

$$K = \frac{E}{3(1 - 2v)} \qquad (C3.6)$$

Table C3.3 Typical values of Poisson's ratios for rocks (after Gercek 2007)

Rock type	v
Andesite	0.20–0.35
Basalt	0.10–0.35
Conglomerate	0.10–0.40
Diabase	0.10–0.28
Diorite	0.20–0.30
Dolerite	0.15–0.35
Dolomite	0.10–0.35
Gneiss	0.10–0.30
Granite	0.10–0.33
Granodiorite	0.15–0.25
Greywacke	0.08–0.23
Limestone	0.10–0.33
Marble	0.15–0.30
Marl	0.13–0.33
Norite	0.20–0.25
Quartzite	0.10–0.33
Rock salt	0.05–0.30
Sandstone	0.05–0.40
Shale	0.05–0.32
Siltstone	0.05–0.35
Tuff	0.10–0.28

and

$$G = \frac{E}{2(1 + v)} \tag{C3.7}$$

Depending on the strain at peak load, the relative ductility of the rock can be classified according to Table C3.4.

Hoek and Brown (1980, 1997) suggest that the uniaxial compressive strength of rock cores are influenced by the core diameter by Equation C3.8:

$$\sigma_{c,d} = \sigma_{c,50} \left(\frac{50}{d} \right)^{0.18} \tag{C3.8}$$

where d = core diameter in mm, $\sigma_{c,d}$ = UCS of a core with a diameter of d mm, and $\sigma_{c,50}$ = UCS of a core with a diameter of 50 mm.

Table C3.4 Relative ductility based on axial
strain at peak load (after Handin 1966)

Classification	Axial strain (%)
Very brittle	< 1
Brittle	1–5
Moderately brittle (transitional)	2–8
Moderately ductile	5–10
Ductile	> 10

Cost: US$80–US$220 (additional US$60 for specimen preparation)

C4   Point Load Test

Objective: To determine the *point load strength index* $I_{s(50)}$

Standards: ASTM D5731
ISRM (1985)*
AS 4133.4.1

Introduction

The *point load test* is an index test for the strength classification of rocks where a piece of rock is held between two conical platens of a portable lightweight tester shown in Figure C4.1. The load is increased to failure and the point load index $I_{s(50)}$ is calculated based on the failure load and the spacing between the cone tips. $I_{s(50)}$ is used to classify the rock and is roughly correlated

(b)

(a)

Figure C4.1 (a) Point load tester (Courtesy of Harrison Meehan, James Cook University, Australia) and (b) conical platen

to the strength parameters such as uniaxial compressive strength σ_c. The test is rather quick and can be conducted on regular rock cores or irregular rock fragments. A key advantage of the point load test is that it can be carried out on an irregular rock fragment; this is not the case with most other tests where the specimens have to be machined and significant preparation is required. This makes it possible to do the tests at the site and on several samples in a relatively short time. Especially during the exploration stage, point load tests are valuable in making informed decisions and can help in selecting the correct samples for the more sophisticated laboratory tests.

The test can also be used to quantify the strength anisotropy $I_{a(50)}$, the ratio of $I_{s(50)}$ in two perpendicular directions (e.g., horizontal and vertical). Historical developments of the point load test and the theoretical background are discussed by Broch and Franklin (1972).

Procedure:

1. The test specimen can be of any of the four forms shown in Figure C4.2 with an equivalent diameter of 30 to 85 mm.

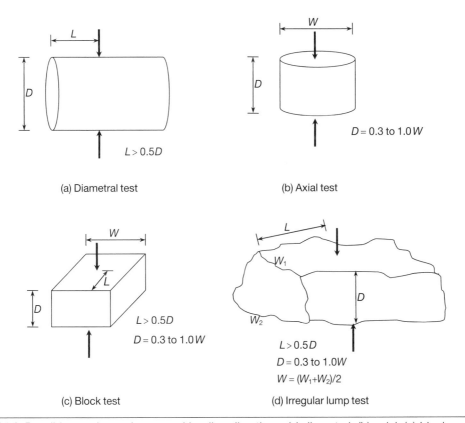

Figure C4.2 Possible specimen shapes and loading directions: (a) diametral; (b) axial; (c) block; and (d) irregular

2. Insert the specimen in the point load tester between the two conical platens, bringing them into contact with the specimen. Note the two conditions: (a) $L > 0.5\ D$ and (b) $D = 0.3$ to $1.0\ W$.
3. Increase the load to failure and record the failure load P, the distance between the platens D, and the failure mode of the specimen. The loading duration should be 10 to 60 s.
4. The test should preferably be carried out on at least 10 samples, and the two highest and lowest values should be discarded. If fewer specimens are tested, the highest and lowest values should be discarded. The mean of the remaining values should be reported as the $I_{s(50)}$ of the rock.
5. The test should be rejected if the failure does not extend to the full depth (D) of the specimen.

Note:

The point load tester should be periodically calibrated using an independently certified load cell and set of displacement blocks, checking the measurements of P and D over the entire range of values expected.

Datasheet:

A simple datasheet for this test, prepared in Excel, is shown in Table C4.1.

The other data recorded and reported may include:

- Source of the sample, project name, and location
- Lithologic (macroscopic) description of the rock
- Information on the water content (at least state whether dry or saturated)
- Apparatus used and model number
- Orientation of the axis of loading with respect to specimen anisotropy (e.g., bedding planes); state whether parallel, perpendicular, or at unknown or random directions
- Tabulation such as in Table C4.1, including the values of P, D, W, D_e, I_s, and $I_{s(50)}$
- Mean value of $I_{s(50)}$; separate values are required for samples tested parallel and perpendicular to the planes of weakness
- The *point load strength anisotropy index* $I_{a(50)}$ defined as the ratio of $I_{s(50)}$ measured perpendicular and parallel to the planes of weakness
- Estimated value of UCS and strength classification
- Photographs of test specimens before and after the test clearly showing the failure mode
- Standard followed and deviations, if any

Analysis:

The *uncorrected point load index* I_s is defined as:

$$I_s = \frac{P}{D_e^2} \tag{C4.1}$$

Table C4.1 Point load test data (after ISRM 1985)

No.	Type	W (mm)	D (mm)	P (kN)	D_e (mm)	I_s (MPa)	$I_{s(50)}$ (MPa)
1	i ⊥	30.4	17.2	2.687	25.8	4.04	~~3.00~~
2	i ⊥	16.0	8.0	0.977	12.8	5.99	3.24
3	i ⊥	19.7	15.6	1.962	19.8	5.01	3.30
4	i ⊥	35.8	18.1	3.641	28.7	4.41	3.44
5	i ⊥	42.5	29.0	6.119	39.6	3.90	3.51
6	i ⊥	42.0	35.0	7.391	43.3	3.95	~~3.70~~
7	b ⊥	44	21	4.600	34.3	3.91	3.30
8	b ⊥	40	30	5.940	39.1	3.89	3.48
9	b ⊥	19.5	15	2.040	19.3	5.48	~~3.57~~
10	b ⊥	33	16	2.870	25.9	4.27	~~3.18~~
11	d //	—	49.93	5.107	49.93	2.05	2.05
12	d //	—	49.88	4.615	49.88	1.85	1.85
13	d //	—	49.82	5.682	49.82	2.29	~~2.29~~
14	d //	—	49.82	4.139	49.82	1.67	~~1.66~~
15	d //	—	49.86	4.546	49.86	1.83	1.83
						0	
16	d //	—	25.23	1.837	25.23	2.89	2.12
17	d //	—	25.00	1.891	25	3.03	2.21
18	d //	—	25.07	2.118	25.07	3.37	~~2.47~~
19	d //	—	25.06	1.454	25.06	2.32	~~1.70~~
20	d //	—	25.04	1.540	25.04	2.46	1.80

a = axial
b = block
d = diametral
i = irregular lump
⊥ = loaded perpendicular to plane of weakness
// = loaded parallel to plane of weakness

Mean $I_{s(50)\perp}$	3.38 MPa
Mean $I_{s(50)//}$	1.98 MPa
$I_{a(50)}$	1.71

where D_e is the equivalent core diameter. In a diametral test (Figure C4.2a), $D_e = D$. In axial, block, or irregular lump tests (Figures C4.2b, C4.2c and C4.2d, respectively), the minimum cross-sectional area of the plane through the platen contact points A is computed as $A = WD$.

Equating this area to that of a circle, the equivalent diameter D_e is computed as:

$$D_e = \sqrt{\frac{4A}{\pi}} = \sqrt{\frac{4WD}{\pi}} \tag{C4.2}$$

I_s increases with D_e, and, therefore, it is desirable to have a unique point load index of the rock sample that can be used in rock strength classification. The *size-corrected* point load strength index $I_{s(50)}$ is defined as the value of I_s obtained if D_e is 50 mm. It can be computed as:

$$I_{s(50)} = I_s \times \left(\frac{D_e}{50}\right)^{0.45} \tag{C4.3}$$

where D_e is in mm.

The ratio of uniaxial compressive strength σ_c to $I_{s(50)}$ can be taken as 20 to 25, but it can vary in the range of 15 to 50 considering extreme possibilities including anisotropic rocks. Bieniawski (1975) suggested that $\sigma_c = 24\ I_{s(50)}$. In spite of the similarities between point load test and Brazilian indirect tensile strength test, any attempt to derive σ_t from $I_{s(50)}$ should be discouraged (Russell and Wood 2009). Classification of rocks based on $I_{s(50)}$ is given in Table C4.2. Point load tests are unreliable for rocks with uniaxial compressive strength less than 25 MPa (Hoek and Brown 1997).

Table C4.2 Rock classification (after AS 1726)

Classification			
Term	Symbol	$I_{s(50)}$ (MPa)	Field description of strength
Extremely low	EL	< 0.03	Easily remolded in hand to a soil-like material.
Very low	VL	0.03 to 0.1	Crumbles under firm blows with sharp end of a pick; can be peeled with a knife; less than 30 mm pieces can be broken by fingers.
Low	L	0.1 to 0.3	Easily scored with a knife; a 50 mm dia × 100 mm long core may be broken by hand; sharp edges of the cores may be friable and break during handling; a firm blow with a pick point gives 1 to 3 mm indentations.
Medium	M	0.3 to 1.0	Readily scored with a knife; a 50 mm dia × 100 mm long core may be broken by hand with difficulty.
High	H	1 to 3	A 50 mm dia × 100 mm long core cannot be broken by hand but can be broken with a pick with a single firm blow.
Very high	VH	3 to 10	Hand specimen breaks after more than one blow with a pick.
Extremely high	EH	>10	Specimen requires many blows with a pick to break through intact material.

Cost: US$30–US$50

C5   Indirect (Brazilian) Tensile Strength

Objective: To determine the tensile strength σ_t of a rock sample

Standards: ASTM D3967
ISRM (1978a)*

Introduction

On rock samples, it is difficult to carry out a direct tensile strength test in the same way we test steel specimens. The main difficulties are in gripping the specimens without damaging them and applying stress concentrations at the loading grip, as well as applying the load without eccentricity. *Indirect tensile strength test*, also known as the *Brazilian test*, is an indirect way of measuring the tensile strength of a cylindrical rock specimen having the shape of a disc. The sample with thickness to diameter (t/d) ratio of 0.5 is subjected to a load that is spread over the entire thickness, applying a uniform vertical line load diametrically (Figure C5.1). The load is increased to failure, where the sample generally splits along the vertical diametrical plane.

From the theory of elasticity of an isotropic medium, the tensile strength of the rock σ_t is given by (Timoshenko 1934; Hondros 1959):

$$\sigma_t = \frac{2P}{\pi dt} \tag{C5.1}$$

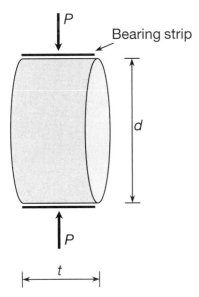

Figure C5.1 Schematic diagram of Brazilian test

where P is the load at failure. The main features of the standard procedure suggested by ISRM (1978) and ASTM D3967 are described briefly. The test works better for brittle materials and has been adopted for concrete, ceramics, cemented soils, and asphalt. Note that the recommended t/d ratio can be different for other materials. Mellor and Hawkes (1971) discussed the test procedure in detail.

Procedure:

1. The test specimen diameter should be at least of NX core size (54 mm), and the thickness should be approximately equal to half the diameter. ASTM D3967 allows t/d ratio of 0.20 to 0.75. The test specimens should be cut and prepared using clean water. The end faces should be flat within 0.25 mm and should be parallel to within 0.25°. The cylindrical surfaces should be free of tool marks and any irregularities should be within 0.025 mm. The diameter of the specimen should be at least 10 times larger than the largest mineral grain present.
2. ISRM (1978a) suggests the loading arrangement shown in Figure C5.2, where the two steel jaws will be in contact with the specimen over an arc length that subtends 10° at the center when failure occurs. It is suggested that the radius of the jaws be 1.5 times the specimen radius. The upper jaw has a spherical seating formed by a 25 mm diameter half ball bearing.
3. Wrap one layer of masking tape around the perimeter of the test specimen.
4. Load the specimen at a constant rate of strain or stress. The measured σ_t is sensitive to the loading rate. The faster the loading, the higher the σ_t (Mellor and Hawkes 1971).

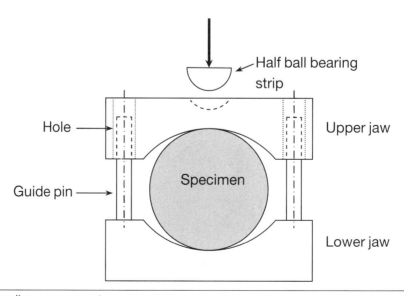

Figure C5.2 Loading arrangement

This phenomenon is commonly seen in soils too. ASTM D3967 suggests that the rate should be selected such that the specimen fails in 1 to 10 minutes.

5. It is not necessary to measure and plot load against displacement to detect the failure load. In the absence of load-displacement plot or measurement, the operator has to be alert and detect the failure load.

6. The number of specimens tested is governed by practical considerations, but generally 10 are recommended. A smaller number is acceptable if the reproducibility of the test results is good (coefficient of variation less than 5%).

Datasheet:

A simple datasheet for this test, prepared in Excel, is shown in Table C5.1.

The other data recorded may include:

- Lithologic (macroscopic) description of the rock
- Orientation of the axis of loading with respect to specimen anisotropy (e.g., bedding planes)
- Sample source, including location, depth, coring technique, storage history, and dates
- Number of specimens tested
- Specimen diameter and height
- Water content at the time of the test
- Test duration and loading rate
- Test date and testing machine details
- Modes of failure
- Tensile strength of each specimen expressed to three significant figures along with the average value
- Standard followed and deviations, if any

Table C5.1 Indirect tensile strength test data

Sample no.	d (mm)	t (mm)	Mass (g)	Volume (cm^3)	Density (g/cm^3)	P (kN)	σ_t (MPa)
BH2_41.2m	63.2	32.1	288.1	100.70	2.86	29	9.1
BH2_41.5m	63.1	31.9	281.3	99.76	2.82	30.8	9.7
BH2_41.9m	63.3	32.1	281.8	101.02	2.79	32.6	10.2
BH2_42.1m	63.2	31.8	282.3	99.76	2.83	29.4	9.3
BH2_42.4m	63.5	31.9	280.9	101.02	2.78	34.4	10.8
BH2_42.7m	63.1	32.1	280.1	100.38	2.79	30.4	9.5
BH3_40.8m	63.4	31.9	283	100.71	2.81	37.7	11.9
BH3_41.2m	63.1	31.8	280.4	99.44	2.82	37.6	11.9
BH3_41.6m	63.6	32.2	283.4	102.30	2.77	35.5	11.0
BH3_41.9m	63.3	31.9	284.1	100.39	2.83	34.2	10.8
Average σ_t (MPa)							10.4

Analysis:

There is little analysis involved in this test where the interpretation is rather straightforward. Tensile strength of a rock is required in most designs, analysis and numerical modeling of underground excavation, tunneling, slope stability, and so forth. Indirect tensile strength can be assumed as approximately equal to the direct tensile strength. In addition, σ_t is assumed to be a fraction of the uniaxial compressive strength (σ_c or q_u). A wide range of values from 1/5 to 1/22 have been reported in the literature, and 1/10 is a good first estimate.

The state of stress at the center of the sample is given by a horizontal tensile stress (σ_t) and a vertical compressive stress that is three times greater in magnitude, both of which are principal stresses.

Cost: US$120–US$150 (additional US$60 for specimen preparation)

C6   Schmidt Hammer Rebound Hardness

Objective: To determine the rebound hardness of a rock core

Standards: ASTM D5873
ISRM (1978c)* and Aydin (2009)*

Introduction

The Schmidt hammer (Figure C6.1) was originally developed in 1948 for testing the hardness of concrete (Schmidt 1948). It is a simple, portable, and inexpensive device that gives the rebound

Figure C6.1 N-type Schmidt hammer (Courtesy of Harrison Meehan, James Cook University, Australia)

hardness value R for an intact rock specimen in the laboratory or the rock mass in situ. The test is generally nondestructive for rocks of at least moderate strength, and therefore the same specimen can be used for other tests. ASTM D5873 and ISRM recommend this test for rocks with UCS of 1 to 100 MPa and 20 to 150 MPa, respectively. The rebound hardness R has been correlated with rock properties such as UCS and E, for example.

The hammer consists of a spring-loaded metal piston that is released when the plunger is pressed against the rock surface. The impact of the piston on the plunger transfers the energy to the rock. How much of this energy is recovered depends on the hardness of the rock and is measured by the rebound height of the piston. The harder the surface, the shorter is the penetration time (i.e., smaller impulse and less energy loss) and the greater the rebound. Rebound hardness R is a number that varies in the range of 0 to 100.

Two types of Schmidt hammers are commonly used. They are the L-type with an impact energy of 0.735 N·m and the N-type with an impact energy of 2.207 N·m. The measured rebound hardness is denoted by R_L and R_N, respectively. Some other notations used in the literature for rebound hardness are H_R, N, and SRH. Prior to 2009, ISRM recommended only L-type hammers; now they are both allowed (Aydin 2009). N-type was mostly used for concrete. However, it is less sensitive to surface irregularities and suits field applications. ASTM does not specify the type of hammer.

Procedure:

1. Calibrate the Schmidt hammer using a calibration test anvil supplied by the manufacturer based on the average of ten readings. A correction factor is suggested as:

$$CF = \frac{\text{Specified standard value of the anvil}}{\text{Average of the 10 readings on the anvil}} \qquad (C6.1)$$

   that has to be applied to all readings. This factor is to account for the spring losing its stiffness with time.

2. Select a representative specimen of at least NX (54 mm) core size with a length greater than 100 mm (ISRM). ASTM suggests a minimum of 150 mm length. For N-type hammers, ISRM suggests 84 mm diameter or larger cores (Aydin 2009).

3. Securely clamp the specimen to a steel base—20 kg for L-type and 40 kg for N-type hammers—located on firm and flat ground. The cylindrical core specimen should be placed in an arc-shaped *cradle* of the same radius machined in the steel base.

4. The hammer should be used in one of three positions: vertically upward, horizontally, or vertically downward with the axis of the hammer ±5° from the desired position.

5. ISRM recommends 20 readings at different locations with an option to stop when any 10 subsequent readings differ only by four. ASTM recommends 10 readings.

6. ISRM (1978c) suggests using the average of the top 10 readings. ASTM recommends discarding the readings that differ from the average by more than 7 and then averaging

the rest. The revised ISRM (Aydin 2009) doesn't suggest discarding any data, and to present the values as a histogram with mean, median, mode, and range. In addition, the revised ISRM suggests normalizing the rebound values with respect to horizontal impact directions and also provides charts.

Datasheet:

A simple datasheet for this test, prepared in Excel, is shown in Table C6.1.

The other data recorded may include:

- Lithologic (macroscopic) description of the rock
- Sample source
- Sample moisture content during the test
- Size and shape of the specimen

Table C6.1 Rebound hardness measurements

	BH1_26.0m		
	Measured	**Corrected**	**Sorted**
Reading 1	50.92	51.71	51.71
Reading 2	56.21	57.08	51.82
Reading 3	51.03	51.82	56.28
Reading 4	55.42	56.28	56.99
Reading 5	58.56	59.47	57.08
Reading 6	62.32	63.28	57.21
Reading 7	59.36	60.28	57.39
Reading 8	56.52	57.39	58.11
Reading 9	57.23	58.11	58.16
Reading 10	58.12	59.02	58.82
Reading 11	59.34	60.26	**59.02**
Reading 12	65.34	66.35	**59.26**
Reading 13	65.23	66.24	**59.47**
Reading 14	56.12	56.99	**59.87**
Reading 15	58.96	59.87	**60.26**
Reading 16	57.27	58.16	**60.28**
Reading 17	56.34	57.21	**60.54**
Reading 18	57.92	58.82	**63.28**
Reading 19	58.36	59.26	**66.24**
Reading 20	59.62	60.54	**66.35**
Average		58.91	**61.46**

Note: CF determined as 1.02

- Orientation of the hammer axis in the test
- Hammer type (L, N, or other)
- Method of clamping the specimen
- Histogram of 20 rebound readings
- Photographs (or description) of the impact points before and after damage
- Standard followed and deviations, if any

Analysis:

In the data shown in Table C6.1, the correction factor (CF) was determined through calibration as 1.02. The corrected values in the third column are obtained by multiplying the measured readings by CF. All 20 values, sorted in ascending order, are given in the fourth column. The average of the top 50% values shown in bold in Table C6.1 (= 61.46) is the rebound hardness as per ISRM (1978c). The revised version suggests presenting all values in the form of a histogram as shown in Figure C6.2.

The average of the 20 readings is 58.91. The four values that differ from this by more than 7 were discarded (shaded in Table C6.1), and the average of the rest was determined as 58.88. This is the rebound hardness suggested by ASTM.

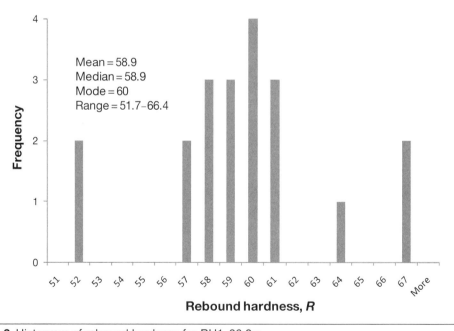

Mean = 58.9
Median = 58.9
Mode = 60
Range = 51.7–66.4

Figure C6.2 Histogram of rebound hardness for BH1_26.0m

Cost: US$50

C7 Slake Durability Test

Objective: To determine the slake durability index of a rock sample consisting of a collection of rock fragments

Standards: ASTM D4644
ISRM (1979b)*
AS 4133.3.4-2005

Introduction

Rocks are generally weaker wet than dry due to the presence of water in the cracks and its subsequent reaction to the applied loads during the tests. Repeated wetting and drying that happens frequently in service can weaken the rock significantly. *Slaking* is a process of disintegration of an aggregate when in contact with water. *Slake durability index* quantifies the resistance of a rock to wetting and drying cycles and is seen as a measure of the *durability* of the rock. This is mainly used for weak rocks such as shales, mudstones, claystones, and siltstones.

Figure C7.1 shows the slake durability apparatus that consists of two rotating sieve mesh drums immersed in a water bath. Ten rock lumps, each weighing 40 to 60 g, are placed in the drum and rotated for 10 minutes, allowing for disintegrated fragments to leave the drum

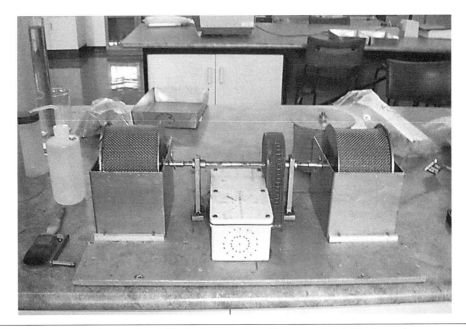

Figure C7.1 Photograph of slake durability test apparatus (Photograph: N. Sivakugan)

through the 2 mm sieve mesh. The remaining fragments in the drum are dried and weighed. This is repeated over a second cycle of slaking, and the dry mass of the sample remaining in the drum, expressed as a percentage of the original mass in the drum at the beginning of the test, is known as the *second-cycle slake durability index* I_{d2} that varies in the range of 0 to 100%. For samples that are highly susceptible to slaking, I_{d2} is close to zero and, for very durable rocks, it is close to 100%.

Procedure:

1. Take a representative sample of 10 rock lumps each with a mass of 40 to 60 g, giving a total mass of 450 to 550 g. The corners should be rounded off so that the lumps are roughly spherical.
2. Place the lumps in a drum and dry them at 105 to 110°C to constant mass and determine the mass m_1 of the drum with the lumps.
3. With the lid in place, mount the drum in the trough, coupled to the motor.
4. Fill the trough with slaking fluid (e.g., tap water at 20°C, sea water, etc. to replicate the service environment) to 20 mm below the drum axis.
5. Rotate the drum that is partially submerged at 20 rev/min for 10 minutes.
6. Remove the lid and place the drum with the remaining sample in the oven to dry at 105 to 110°C. Determine the dry mass m_2 of the drum and the remaining material.
7. Carry out a second cycle of slaking by repeating Steps 3 through 7. Determine the mass m_3 of the drum with the material retained after the second cycle.
8. Brush the dry drum clean and record the mass m_4.
9. Carry out a third or fourth cycle only if required.

Datasheet:

A simple datasheet for this test, prepared in Excel, is shown in Table C7.1.
   The other data recorded may include:

- Lithologic (macroscopic) description of the rock
- Slake durability index I_{d2} to the nearest 0.1%; include I_{d1} when $I_{d2} < 10\%$
- Sample source
- Nature and temperature of the slaking fluid
- Appearance of fragments retained in the drum
- Appearance of material passing through the drum
- Standard followed and deviations, if any

Table C7.1 Slake durability test data

Sample no.	Porcellanite2	Porcellanite7	Claystone1	Claystone3	Claystone8
Mass of drum + dry sample (m_1), g	1476	1457	1464	1493	1503
Mass of drum + dry sample after 1st cycle (m_2), g	1472	1452	1125	1114	1103
Mass of drum + dry sample after 2nd cycle (m_3), g	1467	1446	1013	1004	1009
Mass of drum (m_4), g	971	970	968	969	968
2nd Cycle Slake Durability Index, I_{d2}	**98.2**	**97.7**	**9.1**	**6.7**	**7.7**
1st Cycle Slake Durability Index, I_{d1}	99.2	99.0	**31.7**	**27.7**	**25.2**
Mass of drum + dry sample after 3rd cycle, g (Only if required)	1464	1443	—	—	—
Duration of third cycle (if not 10 minutes)					
3rd Cycle Slake Durability Index, I_{d3}	97.6	97.1	—	—	—
Mass of drum + dry sample after 4th cycle, g (Only if required)	1468.0	1447.0			
Duration of fourth cycle (if not 10 minutes)	30 min	30 min			
4th Cycle Slake Durability Index, I_{d4}	98.4	97.9			
Slaking fluid	Sea water			Tap water	
Temperature of slaking fluid	26°C	26°C	27°C	27°C	27°C

Analysis:

The *first-cycle slake durability index* I_{d1} is defined as:

$$I_{d1} = \frac{m_2 - m_4}{m_1 - m_4} \times 100 \qquad (C7.1)$$

The *second-cycle slake durability index* I_{d2} is defined as:

$$I_{d2} = \frac{m_3 - m_4}{m_1 - m_4} \times 100 \qquad (C7.2)$$

The second-cycle slake durability index I_{d2} is the one that is commonly used as a measure of rock durability. It is recommended to include I_{d1} only in rocks that are classified as very low in durability with $I_{d2} < 10\%$. The durability classification of rocks, based on the slake durability index proposed by Gamble (1971) is given in Table C7.2. This is slightly different from that proposed by Franklin and Chandra (1972). For rocks of higher durability, three or more cycles (i.e., I_{d3}, I_{d4}, etc.) may be appropriate.

Table C7.2 Durability classification based on slake durability index (after Gamble 1971)

Durability	I_{d1}	I_{d2}
Very high	> 99	> 98
High	98–99	95–98
Medium high	95–98	85–95
Medium	85–95	60–85
Low	60–85	30–60
Very low	< 60	< 30

Cost: US$85

C8   Triaxial Test on a Rock Specimen

Objective: To determine the vertical normal stress σ_1 at failure under different confining pressures σ_3, enabling the failure envelope to be developed

Standard: ISRM (1983)*

Introduction

As the first approximation, it can be assumed that rocks, like most geomaterials, follow the Mohr-Coulomb failure criterion given by:

$$\tau_f = c + \sigma \tan \phi \tag{C8.1}$$

where τ_f = shear strength (or shear stress at failure on the failure plane), σ = normal stress on the failure plane, c = cohesion, and ϕ = friction angle. Cohesion and friction angle are the shear-strength parameters of the rock and are constants. Thus, it can be seen from Equation C8.1 that τ_f is proportional to σ. In terms of major and minor principal stresses at failure, Equation C8.1 can be written as:

$$\sigma_1 = \left(\frac{1 + \sin \phi}{1 - \sin \phi} \right) \sigma_3 + 2c \left(\frac{1 + \sin \phi}{1 - \sin \phi} \right)^{0.5} \tag{C8.2}$$

There are also other failure criteria for rocks such as that suggested by Hoek-Brown (1980) where the failure envelope is nonlinear.

Similar to the triaxial tests on soils, here too cylindrical rock specimens are subjected to different confining pressures and loaded axially to failure. The only difference is that the loads and pressures are much higher. The test procedure described, as suggested by ISRM (1983), does not have a provision for pore water pressure or drainage measurements. It is similar to an unconsolidated undrained triaxial test on a clay specimen. Only the procedure for an individual test is described here. The procedures for a multiple failure state test and a continuous failure state test, similar to a stages test, are given in ISRM (1983).

Procedure:

1. The test specimen diameter should be at least of NX core size (54 mm), and the length should be approximately equal to 2.0 to 3.0 times the diameter. The test specimens should be cut and prepared using clean water. The ends of the test specimens shall be flat to ± 0.01 mm and be parallel to each other and at right angles to the longitudinal axis. The sides of the specimens shall be smooth and free of abrupt irregularities and straight to within 0.3 mm over the full length of the specimen. The diameter of the specimen should be at least 10 times larger than the largest mineral grain present. Use of capping material or end surface treatment is not permitted.

2. Take two measurements each of the diameter at right angles at top, middle, and bottom of the core to nearest 0.1 mm. The average of the six readings should be used as the original diameter d_0 of the specimen.
3. Determine the original height of the specimen l_0 to the nearest 1.0 mm.
4. Enclose the specimen in a flexible membrane to prevent the confining fluid from entering the specimen pores (Figure C8.1a). Sometimes it is required to get the customized membranes made to suit the different core diameters.
5. Assemble the specimen, platens, and the measurement devices.
6. Fill the triaxial cell with oil that is used as the confining fluid.
7. Apply the confining pressure and record the value.
8. Maintaining the constant cell pressure, apply the vertical stress on the sample until failure occurs (Figure C8.1b). Apply vertical load at a constant strain/stress rate (e.g., 0.5 to 1.0 MPa/s) such that the specimen fails in 5 to 15 minutes.

Datasheet:

A simple datasheet for this test, prepared in Excel, is shown in Table C8.1.

The other data recorded may include:

- Lithologic (macroscopic) description of the rock
- Sample source
- Standard followed and deviations, if any

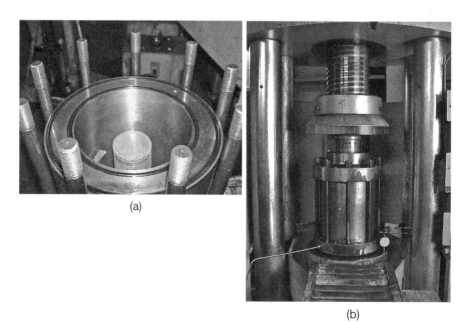

(a)

(b)

Figure C8.1 Rock triaxial test: (a) cell with specimen enclosed within a membrane and (b) while loading (Photograph: N. Sivakugan)

Table C8.1 Rock triaxial test datasheet

Sample no.	Diameter (mm)	Length (mm)	Mass (g)	Unit wt (kN/m³)	Cell pressure (MPa)	Additional vertical stress at failure (MPa)	σ_3 (MPa)	σ_1 (MPa)
JH_23	63.4	129	1150	27.7	5	95.5	5	100.5
JH_32	63.3	130	1181	28.4	25	134.5	25	159.5
JH_41	63.2	130	1174	28.3	50	157.0	50	207.0

Analysis:

Three or more triaxial tests, at different confining pressures, are required to define the failure envelope. By drawing Mohr circles or plotting σ_1 against σ_3, the failure envelope can be defined and the shear strength parameters can be determined. For intact rocks, Hoek-Brown (1980) strength criterion suggests that the effective stresses σ_1' and σ_3' at failure are related by:

$$\sigma_1' = \sigma_3' + \sigma_c \left(m_i \frac{\sigma_3'}{\sigma_c} + 1 \right)^{0.5} \tag{C8.3}$$

where σ_c is the uniaxial compressive strength of the intact rock specimen, and m_i is a rock constant that varies between 4 and 35. Selected m_i values are given in Table C8.2. In the absence of σ_c values from a uniaxial compression test, estimates from Table C8.2 can be used.

Table C8.2 m_i values for rocks (after Hoek and Brown 1997; Wyllie and Mah 2004)

		Texture		
	Coarse	**Medium**	**Fine**	**Very fine**
Sedimentary	Conglomerates (21 ± 3) Breccias (19 ± 5)	Sandstones 17 ± 4	Siltstones 7 ± 2 Greywackes (18 ± 3)	Claystones 4 ± 2 Shales (6 ± 2) Marls (7 ± 2)
	Crystalline limestone (12 ± 3)	Sparitic limestone (10 ± 2)	Micritic limestone (9 ± 2)	Dolomite (9 ± 3)
		Gypsom 8 ± 2	Anhydrite 12 ± 2	Chalk 7 ± 2
Metamorphic	Marble 9 ± 3	Hornfels (19 ± 4) Metasandstone (19 ± 3)	Quartzite 20 ± 3	
	Migamatite (29 ± 3)	Amphibiolites 26 ± 6	Gneiss 28 ± 5	
		Schists 12 ± 3	Phyllites (7 ± 3)	Slates 7 ± 4
Igneous	Granite (32 ± 3) Granodiorite (29 ± 3)	Diorite (25 ± 5)		
	Gabro 27 ± 3	Dolerite (16 ± 5)	Diabase (15 ± 5)	Peridotite (25 ± 5)
	Norite 20 ± 5	Rhyolite (25 ± 5)	Dacite (25 ± 3)	Obsidian (19 ± 3)
	Porphyries (20 ± 5)	Andesite 25 ± 5	Basalt (25 ± 5)	
	Agglomerate (19 ± 3)	Breccia (19 ± 5)	Tuff (13 ± 5)	

Note: The values in parenthesis are estimates. The others are measured.

In the laboratory triaxial test on rocks, the cell pressure is σ_3, and the vertical stress at failure is σ_1. When there are no pore water pressures, they are the same as the effective stresses used in Equation C8.3. Normalizing with σ_c, Equation C8.3 can be written as:

$$\frac{\sigma_1'}{\sigma_c} = \frac{\sigma_3'}{\sigma_c} + \left(m_i \frac{\sigma_3'}{\sigma_c} + 1 \right)^{0.5}$$

The experimental data can be compared against the plot of $\dfrac{\sigma_1'}{\sigma_c}$ versus $\dfrac{\sigma_3'}{\sigma_c}$ for different values of m_i. Hoek and Brown (1997) recommend selecting confining pressure such that $\dfrac{\sigma_3'}{\sigma_c} < 0.5$ so that the values of m_i can be compared against their estimates.

Cost: US$350

C9   Direct Tensile Strength

Objective: To determine the tensile strength σ_t of a rock sample

Standards: ASTM D2936
ISRM (1978a)*

Introduction

Rocks are weaker in tension than in compression. Unlike most soils, they do have significant tensile strength. In rock designs and modeling, the tensile strength of the rock is always one of the major input parameters. On rock samples it is difficult to carry out a direct tensile strength test in the same way we test steel specimens. The main difficulties are in gripping the specimens without damaging them and applying stress concentrations at the loading grip as well as applying the load without eccentricity. A better method to use is to glue the specimen to metal ends and then pull them apart.

Procedure:

1. The test specimen diameter should be at least of NX core size (54 mm), and the length should be approximately 2.5 to 3.0 times the diameter. The ends should be smooth and flat and perpendicular to the axis of the specimen within 0.001 radians. The sides of the specimen should be smooth and free of any irregularities and straight to within 0.1 mm over the entire length.
2. Measure the diameter of the specimen to the nearest 0.1 mm by averaging two diameters measured at right angles to each other at mid-height.
3. Measure the height of the specimen to the nearest 1.0 mm.
4. Cement a metal cap to each flat end of the specimen. The cap should be at least 15 mm thick. It should be larger in diameter than the specimen, but the difference should be less than 2 mm. The caps should be provided with a suitable linkage system for load transfer from the loading device to the test specimen without applying bending or torsional stresses. The length of the linkages at each end should be at least twice the diameter of the metal caps. The thickness of the cement layer should be less than 1.5 mm at each end.
5. After the cement has hardened sufficiently, with the bond strength in excess of the tensile strength of the rock, place the sample and properly align the load transfer system in the testing machine.
6. Apply the load at a constant rate such that the specimen fails within 5 minutes of loading. Alternatively use stress rate of 0.5 to 1.0 MPa/s.

7. Record the maximum load on the specimen to within 1%.
8. The test should be carried out on at least 5 specimens.

Datasheet:

There is no datasheet for this test. The following data are recorded:

- Lithologic (macroscopic) description of the rock
- Orientation of the axis of loading with respect to specimen anisotropy (e.g., bedding planes)
- Sample source, including location, depth, coring technique, storage history, and dates
- Number of specimens tested
- Specimen diameter and height
- Water content at the time of the test
- Test duration and loading rate
- Test date and testing machine details
- Modes of failure
- Tensile strength of each specimen expressed to three significant figures along with the average value
- Standard followed and deviations, if any

Analysis:

There is little analysis involved in this test. The average diameter measured and the maximum load is used in computing the tensile strength.

Cost: US$250 (additional US$60 for specimen preparation)

 Web
Added
Value™

This book has free material available for download from the
Web Added Value™ resource center at *www.jrosspub.com*

Aggregate Testing

Part D

INTRODUCTION

The test procedures for aggregates described in this section are based on Australian Standards (AS), ASTM International Standards (ASTM), and British Standards (BS). The procedure of a specific test closely follows the standard shown with an asterisk.

Aggregates used in engineering applications are mostly collected from quarries supplying virgin materials or stockpiles at recycling yards. Samples of aggregates are usually collected in large plastic sampling bags that can store samples 10 to 15 kg. The sampling bags are labeled with sampling site, type of sample, class of sample, and sampling date when they are collected. This enables easy identification of the samples when they are brought to the laboratory.

Most of the tests described in this section on aggregate testing are based on laboratory testing of recycled aggregates undertaken by the authors in Australia. Tests carried out to date on recycled aggregates by the authors include testing of crushed concrete (20 mm), crushed rock (20 mm), crushed glass (5 mm), and recycled asphalt (20 mm), particularly for pavement sub-base applications. Figures D.1 to D.4 show stockpiles of crushed bricks, glass, concrete, and

Figure D.1 Crushed brick aggregate stockpile (Courtesy of Alex Fraser Recycling, Victoria, Australia)

Figure D.2 Crushed glass aggregate stockpile (Courtesy of Alex Fraser Recycling, Victoria, Australia)

Figure D.3 Crushed concrete aggregate stockpile (Courtesy of Alex Fraser Recycling, Victoria, Australia)

Figure D.4 Crushed rock aggregate stockpile (Courtesy of Alex Fraser Recycling, Victoria, Australia)

rocks at a recycling site. With the scarcity of virgin geomaterials, there is an increasing tendency to use recycled aggregates in engineering applications such as road work. This also serves as a major step toward sustainability in minimizing the waste that has to be disposed.

Aggregates are often blended with other aggregates as part of a mix or blend design. For example, recycled crushed rock or concrete are often blended with varying percentages of recycled crushed glass or brick in projects where recycled blends are used in pavement and footpath sub-base designs. Up to 50% blends with other aggregates are used in some mix designs.

Common laboratory tests for aggregates include geotechnical and pavement-related tests to determine the physical, mechanical, and strength properties of the aggregates. Chemical tests are sometimes undertaken to determine the risk assessment for contaminants in aggregates.

This book provides a detailed description for the physical, mechanical, and strength tests for aggregates. A brief overview is also provided for chemical testing of aggregates.

D1   Water Absorption of Aggregates

Objective: To determine the water absorption of coarse and fine aggregates

Standards: ASTM C127 & C128
AS 1141.5 & 1141.6.1

Introduction

Aggregates can have cracks on the surface that may be connected to the internal pores. When in contact with water, these aggregates can absorb some water that leads to an internal water content while the surface is dry. Water absorption is defined as the ratio of the mass of water held in these internal voids of the aggregates to their oven-dry mass, expressed as a percentage. This is generally in the order of 0 to 5%. This test is required for aggregates used in roadwork and concrete. The mass of the water filling the internal voids is determined by soaking the aggregates in the water for 24 hours and determining their *saturated surface-dry* mass. Subtracting the dry mass of the aggregates gives the mass of water within the internal voids.

In soils containing coarse and fine aggregates, the water absorption is determined separately (W_a and W_A), and a weighted average value (WA_A) is computed based on their relative proportions. The cut-off between fine and coarse is 4.75 mm (No. 4 sieve size) in aggregates; in geotechnical engineering context it is mostly 0.075 mm (No. 200 sieve size).

Procedure:

Fine aggregates:

1. The test portion shall be prepared as follows:
    a) Obtain a sample of aggregates where the fine fraction (smaller than 4.75 mm) is at least 500 g.
    b) If more than 10% of the material is retained on a 4.75 mm sieve, this fraction shall be tested separately using the procedure for coarse aggregates. Ignore the material retained on 4.75 mm sieve if it is less than 10%.
    c) Obtain a test portion of at least 500 g by riffling the sample sieved in Step (b).
2. Immerse the test portion in water at room temperature for a period of not less than 24 hours. Remove the air entrapped in the aggregates by gentle agitation with a rod until no air bubbles rise to the surface.
3. Drain the water off the test portion and spread the aggregates on a flat impervious surface.
4. Surface dry the aggregates by exposing them to a gently moving current of warm air and stirring it frequently to achieve uniform drying.

5. When the aggregates appear to be free flowing, fill the conical mold (Figure D1.1a) by loosely placing part of the test portion in it. Tamp the surface of the aggregates with the tamping tool (Figure D1.1b) 25 times, allowing the tamping tool to fall from about 10 mm above the surface of the aggregates.

(a)

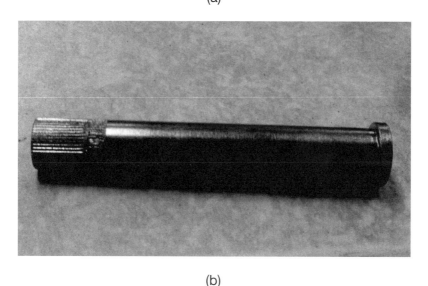

(b)

Figure D1.1 Apparatus for water absorption of fine aggregates: (a) conical mold and (b) tamping tool (Photograph: N. Sivakugan)

6. Lift the conical mold vertically. If free moisture is present, the cone of fine aggregates will retain its shape. If the cone slumps on removal of the mold the first time, the aggregates are too dry and additional water will need to be added and the test portion allowed to stand for 30 minutes.
7. Continue drying with constant stirring and retest at frequent intervals using the procedure in Steps 5 and 6 until the cone of aggregates slumps on the removal of the mold. Slumping of the aggregate cone indicates that it has reached a saturated surface-dry condition.
8. Immediately after the saturated surface-dry condition has been achieved, without any free moisture present, determine the mass of the total test portion, m_2.
9. Dry the aggregates in an oven at 105 to 110°C to constant mass. Determine the mass of dry aggregates, m_1.

Coarse aggregates:

1. Test portions shall be prepared as follows:
    a) Obtain a sample of sufficient mass to permit the preparation of a test portion retained on the 4.75 mm sieve of at least 2 kg for sizes up to and including 20 mm aggregate and at least 5 kg for larger sizes.
    b) Sieve the sample over a 4.75 mm sieve. Reject the undersize material if it amounts to less than 10% of the total. If the amount of undersize material amounts to more than 10%, test it separately. Wash the sample thoroughly to remove dust or other coatings from the surface of the particles.
    c) From the washed material obtain a test portion of at least 2 kg for sizes up to and including 20 mm aggregate, or 5 kg for larger sizes.
2. Immerse the material in water at room temperature with a cover of at least 20 mm of water above the top of the material for a period of at least 24 hours. Stir the material occasionally to dislodge air bubbles. The material remains completely immersed during the soaking period. Materials with water absorption greater than 5% should be soaked for 24 to 26 hours.
3. Following immersion, surface dry the material. Large stones can be dried individually while the finer material may be rolled on a dry cloth. Spread the material one stone deep over a dry cloth and allow it to surface dry, turning the stone at least once during this period. A gentle current of air may be used to accelerate the drying. Drying continues until all visible films of water have been removed but the surfaces of particles still appear damp. Determine the mass of the surface dry material, m_2.
4. Dry the material in the oven at 105 to 110°C to constant mass and determine its dry mass, m_1.

Datasheet:

A simple datasheet for this test, prepared in Excel, is shown in Table D1.1. The water absorption should be present to the nearest 0.1%.

The other data recorded may include:

- Sample source
- Sampling date
- Standard followed and deviations, if any

Analysis:

Fine aggregates:

The water absorption (W_a) of the fine aggregate is computed from:

$$W_a(\%) = \frac{m_2 - m_1}{m_1} \times 100 \tag{D1.1}$$

where m_1 = dry mass of the sample, m_2 = saturated surface-dry mass of the sample, and W_a = water absorption of the fines.

Table D1.1 Water absorption test datasheet

Soil description:	Clayey sandy gravel			
Sample location:	Clayton campus			
Date: 21 May 2010		Test procedure: AS 1141.5 & 1141.6.1		
Notes: Three samples tested from the same bag				

% of fine and coarse fractions:		Sample no.		
	1	2	3	4
% fines (smaller than 4.75 mm)	25	25	25	
% coarse (larger than 4.75 mm)	75	75	75	

Fines:		Sample no.		
	1	2	3	4
Mass of dry sample, m_1 (g)	485	480	499	
Mass of saturated surface-dry sample, m_2 (g)	510	505	525	
Water absorption, W_a (%)	5.2	5.2	5.2	

Coarse:		Sample no.		
	1	2	3	4
Mass of dry sample, m_1 (g)	2088	2085	1987	
Mass of saturated surface-dry sample, m_2 (g)	2155	2150	2050	
Water absorption, W_A (%)	3.2	3.1	3.2	
Weighted average water absorption, WA_A (%)	3.7	3.6	3.7	

Coarse aggregates:

The water absorption (W_A) of the coarse aggregate is computed from:

$$W_A (\%) = \frac{m_2 - m_1}{m_1} \times 100 \tag{D1.2}$$

where m_1 = dry mass of the sample, m_2 = saturated surface-dry mass of the sample, and W_a = water absorption of the coarse aggregates.

When the aggregates contain coarse and fine fractions, they are tested separately and their respective water absorption values (W_a and W_A) and their relative proportions are used in determining a weighted average value (WA_A):

$$WA_A (\%) = \frac{W_a \times \%\text{fines} + W_A \times \%\text{coarse}}{100} \tag{D1.3}$$

Cost: US$80–US$100

D2   Flakiness Index

Objective: To determine the flakiness index of coarse aggregates

Standards: BS 812–105.1:1989*
AS 1141.15–1999

Introduction

Flakiness is one of the measures of the particle shape that plays a key role in the packing density in a soil matrix. Testing for flakiness is important when dealing with coarse aggregates. A particle is classified as flaky when the thickness is less than 60% of the mean size. The mean size is simply the mean of the smallest sieve size through which the particle passes and the largest sieve size on which it is retained. The flakiness index is the percentage of the flaky aggregates within the entire aggregates. Special *slotted sieves* or *thickness gage* with elongated holes can be used to identify the flaky particles (Figure D2.1). BS 812-105.1:1989 suggests that the test be limited to soil grains of 6.3 mm to 63.0 mm.

Figure D2.1 Slotted sieves with thickness gage for flakiness test (Photograph: N. Sivakugan)

Procedure:

The procedure described is based on BS 812-105:1989. The sieve sizes and corresponding slot widths (Figure D2.1) can be slightly different depending on the country and standards used.

1. Reduce the sample by quartering or splitting to produce a test portion that complies with Table D2.1. Dry the test portion by heating at a temperature of 105 ± 5°C to achieve a dry mass that is constant to within 0.1 %.

    Table D2.1 Minimum mass of test portion
    (BS 812-105.1:1989)

Nominal size of material (mm)	Minimum mass of test portion after rejection of oversize and undersize particles (kg)
50	35
40	15
28	5
20	2
14	1
10	0.5

2. Carry out a sieve analysis by using the set of standard sieves (with square holes) suggested in Table D2.2. The suggested aperture sizes and the matching slot widths (Tables D2.2 and D2.3) are slightly different for other standards such as AS 1141.15.

    Table D2.2 Particulars of sieves (BS 812-105.1:1989)

Nominal aperture sizes (mm) Square hole perforated plate 450 mm or 300 mm diameter
63.0
50.0
37.5
28.0
20.0
14.0
10.0
6.30

3. Record the mass m_1 of each size-fractions retained on the sieves (other than the 63.0 mm test sieve) to the nearest 1 g and store them in separate trays with their size marked on the trays. The total mass in consideration is Σm_1. Where the mass of any size-fraction is

considered to be excessive, the fraction may be subdivided provided that the mass of the subdivided fraction is not less than half the appropriate mass given in Table D2.3.

Table D2.3 Aperture sizes and corresponding slot thicknesses with minimum mass for subdivision (BS 812-105.1:1989)

BS test sieve nominal aperture size (mm)		Slot width in thickness gage or slotted sieve (mm)	Minimum mass for subdivision (kg)
100 % passing	100 % retained		
63.0	50.0	33.9 ± 0.3	50
50.0	37.5	26.3 ± 0.3	35
37.5	28.0	19.7 ± 0.3	15
28.0	20.0	14.4 ± 0.15	5
20.0	14.0	10.2 ± 0.15	2
14.0	10.0	7.2 ± 0.1	1
10.0	6.30	4.9 ± 0.1	0.5

Note: Slot width in column 3 is 60% of the mean of the first two columns

4. From the values of m_1, % passing or retained corresponding to the different sieve sizes can be determined. Discard any size-fraction where the mass is less than 5% of the total mass Σm_1, and use this new total mass as Σm_1 in computing the flakiness index.
5. Now, it is required to separate the flaky particles in each of the size-fractions in Step 3. This can be done in one of the following two ways:
   a. *Using the special slotted sieves* (Figure D2.1): Select the special slotted sieve appropriate to the size-fraction being tested. Place the whole of the size-fraction into the sieve and shake until the majority of the flaky particles have passed through the slots. Then gage the particles retained by hand.
   b. *Using the thickness gage* (Figure D2.1): Select the thickness gage slot appropriate to the size-fraction being tested and gage each particle of that size-fraction separately by hand.
   Record the mass of particles m_2 passing each slot size (and thus flaky). The total mass of flaky particles is Σm_2. Note that the slot width is 60% of the mean aperture of the sizes corresponding to the size-fraction being tested (see Table D2.3).

Datasheet:

A simple datasheet for this test, prepared in Excel, is shown in Table D2.4.
   The other data recorded may include:

- Sample source
- Sampling date
- Standard followed and deviations, if any

Table D2.4 Flakiness index test datasheet

Soil description: Sandy gravel

Sample location: Clayton campus

Date: 21 May 2010   Test procedure: BS 812-105.1:1989

Notes:

Standard sieve			Flakiness (slotted) sieve		
Sieve size (mm)	Mass retained (g)	% retained	Width of slot (mm)	Mass passing (g)	% of flaky particles
50.0	0.0	0.0	33.9	0.0	0.0
37.5	0.0	0.0	26.3	0.0	0.0
28.0	0.0	0.0	19.7	0.0	0.0
20.0	0.0	0.0	14.4	0.0	0.0
14.0	2517.8	32.1	10.2	331.2	13.2
10.0	2430.7	31.0	7.2	315.6	13.0
6.3	2887.4	36.8	4.9	601.3	20.8
Total	7835.9			1248.1	
Total flakiness index		16			

Analysis:

The flakiness index is calculated to the nearest whole number as:

$$\text{Flakiness index} = \frac{\Sigma m_2}{\Sigma m_1} \times 100 \tag{D2.1}$$

The flakiness index for each size fraction can be calculated as $(m_2/m_1) \times 100$.

Cost: US$100–US$120

D3   Fines Content

Objective: To determine the fines content of aggregates

Standard: AS 1141.12*

Introduction

The fines content in aggregates refers to the amount of fines (material finer than 75 (m) that has to be determined by washing, also known as wet sieving. Clay particles and other aggregate particles that are dispersed by the wash water and water soluble materials will be removed from the aggregate during the test (AS 1141.12-1996).

Procedure:

1. The minimum mass of aggregates to be used in the test should be based on Table D3.1.

Table D3.1 Minimum mass of test portion for sieving (AS 1141.12:1996)

Nominal size (mm)	75	40	28	20	14	10	7	5	Fine aggregate	Fillers
Graded aggregate	30 kg	15 kg	5 kg	3 kg	1.5 kg	800 g	500 g	300 g	150 g	25 g
One-sized aggregate	25 kg	10 kg	4 kg	1.5 kg	700 g	500 g	300 g	200 g	100 g	—

Note: In Table D3.1, decide whether the aggregate is *graded* or *one-sized* from the percentages retained in the stack of sieves consisting of 75.0 mm, 37.5 mm, 26.5 mm, 19.0 mm, 13.2 mm, 9.50 mm, 6.70 mm, 4.75 mm, 2.36 mm, 0.600 mm, and 0.075 mm. If more than 15% is retained on at least three consecutive sieves, it is a graded aggregate. If more than 60% passes a sieve and is retained by the subsequent sieve, it is a one-sized aggregate.

2. Dry the test portion to constant mass m_1 and record the value.
3. Place the dried test portion in a dish and add clean potable water to cover it. For materials containing clay, a dispersing agent that does not react with the test portion may be used.
4. Vigorously agitate the contents without spilling the water to take the fine material into suspension and immediately pour the wash water through the pair of sieves nested together with the 1.18 mm sieve at the top and 75 μm sieve at the bottom, as shown in Figure D3.1. Avoid decantation of the coarse particles of the test portion. Return the material retained on the sieves to the test portions.
5. Continue agitation and washing until the water passing through the sieves and in the pan is clear.

Figure D3.1 1.18 mm and 75 μm sieves for wet sieving (Photograph: N. Sivakugan)

6. Dry the washed aggregate to constant mass and determine the mass m_2 of the dried aggregate.

Datasheet:

A simple datasheet for this test, prepared in Excel, is shown in Table D3.2. A percentage finer than 75 μm should be reported to the nearest whole number.

The other data recorded may include:

- Sample source and description
- Sampling date
- Method of drying if other than by the oven at 105 to 110°C
- Standard followed and deviations, if any

Analysis:

The percentage passing the 75 μm sieve can be calculated as:

$$\% \text{ passing } 75\mu m = \frac{m_1 - m_2}{m_1} \times 100 \qquad \text{(D3.1)}$$

Table D3.2 Fines content datasheet

Soil description: Clayey sandy gravel
Sample location: Swinburne campus
Date: 17 June 2010   Test procedure: AS 1141.12
Notes: Five samples from bag A

	Test 1	Test 2	Test 3	Test 4	Test 5	Test 6
Sample No.	A 43	A 44	A45	A46	A47	
Dry masses of the test portions:						
Before washing, m_1 (g)	3000	3005	3006	3004	3008	
After washing, m_2 (g)	2720	2700	2705	2710	2715	
% finer than 75 µm	9.33	10.15	10.01	9.79	9.74	

where

m_1 = the mass of the dried test portion before washing, in grams
m_2 = the mass of the dried washed test portion, in grams

Cost: US$40–US$50

D4   Aggregate Impact Value (AIV)

Objective: To determine the AIV that gives a relative measure of the resistance of an aggregate to a sudden shock or impact

Standard: BS 812-112: 1990*

Introduction

Due to repeated vehicle movements, road aggregates are subjected to regular impact loads. The toughness of the aggregates to resist such impact loading is quantified through the aggregate impact test. This test uses hammer blows to produce impact loading on an aggregate sample contained in a 102 mm diameter and 50 mm tall cup (Figure D4.1). The percentage of the sample that passes a 2.36 mm sieve after 15 blows from a 14 kg hammer dropped 380 mm is defined as the AIV. For base course aggregates, it is typically in the range of 10 to 30.

This test is conducted on 10 to 14 mm aggregates in dry and soaked conditions separately. A test specimen is prepared in a 75 mm diameter and 50 mm high cylindrical metal measure,

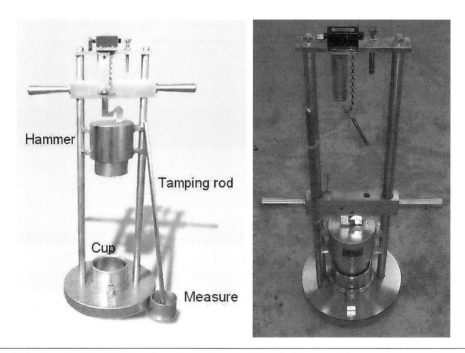

Figure D4.1 Aggregate impact value apparatus (Photograph: N. Sivakugan)

which is simply used as a measure to prepare the same sample mass for two or more subsequent tests. This sample is transferred to the cup where it is subjected to the impact load.

Procedure:

Reduce the laboratory sample by quartering or splitting to produce a test portion of sufficient mass to produce three test specimens of 14 mm to 10 mm size fraction (Table D4.1). Conduct the test by either the dry condition or soaked condition described below.

Table D4.1 Guide to minimum mass of test portions required to obtain a suitable mass of material to determine aggregate impact value (BS 812-112:1990)

Grade of the aggregate (mm)	Minimum mass of the test portion[a] (kg)
All-in aggregate 40 max. size	20
All-in aggregate 20 max. size	15
Graded aggregate 40 to 5	12
Graded aggregate 20 to 5	8
Graded aggregate 14 to 5	5

[a] For normal density aggregates.

Dry condition:

1. Thoroughly sieve the entire dried test portion on the 14 mm and 10 mm test sieves to remove the fraction outside 10 to 14 mm range. Divide the 10 to 14 mm size fraction to produce three test specimens, each of sufficient mass to fill the 75 mm diameter measure.
2. Dry the test specimens by heating at a temperature of 105 ± 5°C for at least 4 hours. Cool to room temperature before testing.
3. Fill the measure to overflowing with the aggregate using a scoop. Tamp the aggregate with 25 blows of the rounded end of the tamping rod; each blow consists of allowing the tamping rod to fall freely from a height of about 50 mm above the surface of the aggregate and the blows being evenly distributed over the surface.
4. Using the tamping rod as a straight edge strike off the surplus aggregate. Fill any obvious depressions with added aggregate. Record the net mass of aggregate in the measure and use the same mass for the other test specimens.
5. Rest the impact machine, without wedging or packing, on the level plate, a block, or the floor so that it is rigid and the hammer guide columns are vertical. Before fixing the 102 mm diameter cup to the impact machine, transfer the whole test specimen from the measure to the cup and then compact by 25 strokes of the tamping rod. With minimal disturbance to the test specimen, fix the cup firmly in position on the base of the machine. Adjust the height of the hammer so that its lower face is 380 ± 5 mm above the upper surface of the aggregate in the cup and then allow it to fall freely onto

the aggregate. Subject the test specimen to a total of 15 such blows, each being delivered at an interval of not less than 1 second.

6. Remove the crushed aggregate by holding the cup over a clean tray and hammering on the outside with a rubber mallet until the particles are sufficiently disturbed to enable the specimen to fall freely onto the tray. Transfer the fine particles adhering to the inside of the cup and the underside of the hammer to the tray by means of a stiff bristle brush.

7. Determine the mass m_1 of aggregate used in the test to the nearest 0.1 g.

8. Sieve the entire specimen in the tray on the 2.36 mm test sieve until no further significant amount passes during a further period of 1 minute. Determine the masses of the fractions passing and retained on the sieve to the nearest 0.1 g (m_2 and m_3 respectively). If the total mass $m_2 + m_3$ differs from the initial mass m_1 by more than 1 g, discard the result and test another specimen.

9. Repeat the procedure as described using a second specimen of the same mass as the first specimen.

Soaked condition:

1. Place each test specimen in the wire basket and immerse it in water in a container with a cover of at least 50 mm of water above the top of the basket. Immediately after immersion, remove the entrapped air from the specimen by lifting the basket 25 mm above the base of the container and allowing it to drop 25 times at a rate of about once a second. Keep the basket and aggregate completely immersed during the operation and for a subsequent period of 24 ± 2 hours and maintain the water temperature at 20 ± 5°C.

2. After soaking, remove the specimen of aggregate from the basket and blot the free water from the surface with the absorbent cloths. Carry out the preparation and testing immediately after this operation.

3. Follow the test procedure described for dry condition (Steps 1 through 9 above), except that the number of blows of the hammer to which the aggregate is subjected is the number of blows that will give between 5 and 20% of fraction smaller than 2.36 mm.

4. Remove the crushed specimen from the cup and dry it in the oven at a temperature of 105 ± 5°C either to constant mass or for at least 12 hours. Allow the dried material to cool and weigh to the nearest gram and record the mass of the test specimen m_1, and the mass m_2 of the fraction passing 2.36 mm sieve.

Analysis:

Aggregate in the dry condition:

For each test specimen, the AIV is expressed as:

$$AIV\ (\%) = \frac{m_2}{m_1} \times 100 \tag{D4.1}$$

where

m_1 = the mass of the test specimen
m_2 = the mass of the material passing the 2.36 mm test sieve

Aggregate in the soaked condition:

For each test specimen, the AIV is expressed as:

$$AIV\ (\%) = \frac{m_2}{m_1} \times \frac{15}{n} \times 100 \qquad \text{(D4.2)}$$

Table D4.2 Aggregate impact value datasheet

Soil description: Gravel
Sample location: CSR quarry
Date: 21 May 2010   Test procedure: BS 812-112: 1990
Notes: Sample B

Aggregate impact value—Aggregate in the dry condition					
Sample no.	1	2	3	4	5
m_1	627.3	628.5			
m_2	62.2	63.2			
m_3	564.5	564.8			
$m_2 + m_3$	626.7	628.0			
AIV	9.9	10.1			
Average AIV			10.0		

m_1 = the mass of the test specimen (g)
m_2 = mass of the material passing the 2.36 mm test sieve (g)
m_3 = mass of the material retained on 2.36 mm test sieve (g)
AIV = the aggregate impact value, as a percentage

Aggregate impact value—Aggregate in the soaked condition					
Sample no.	1	2	3	4	5
m_1	631.2	629.4			
m_2	59.3	64.3			
m_3	571.3	564.6			
$m_2 + m_3$	630.6	628.9			
n	11	11			
AIV	12.8	13.9			
Average AIV			13.4		

m_1 = the mass of the oven-dried test specimen (g)
m_2 = the mass of the oven-dried material passing the 2.36 mm test sieve (g)
m_3 = mass of the oven-dried material retained on 2.36 mm test sieve (g)
n = the number of hammer blows to which the specimen is subjected
AIV = the aggregate impact value, as a percentage

where

m_1 = mass of the oven-dried specimen
m_2 = mass of the oven-dried fraction passing 2.36 mm sieve
n = number of hammer blows

Datasheet:

A simple datasheet for this test, prepared in Excel, is shown in Table D4.2. The AIV, expressed as a percentage, should be reported to the first decimal place.

The other data recorded may include:

- Sample source
- Sampling date
- Standard followed and deviations, if any

Cost: US$100–US$120

D5   California Bearing Ratio (CBR)

Objective: To determine the California Bearing Ratio (CBR) of a *laboratory compacted* soil that is typically used as subgrade, sub-base, or base of a pavement

Standards: AS 1289.6.1.1
BS 1377-4
ASTM D1883-07

Introduction

The *California Bearing Ratio (CBR)* test is an empirical test developed in the United States in the 1930s in California for estimating the bearing values of materials for subgrades, sub-bases, and bases that were used in highways and airfields. The test is performed in the laboratory by pushing a 49.63 mm diameter standard plunger into the soil contained in a 152 mm diameter cylindrical mold at a fixed rate of penetration (1 mm/min) and continuously measuring the force required to maintain that rate. From the resulting load-penetration relationship, the CBR value can be derived. The load-penetration curve is compared against that of a *compacted crushed Californian limestone rock* that is defined to have a CBR of 100. The standard load-penetration values for the compacted crushed rock, against which the comparison is made, are given in Table D5.1. The comparison is made specifically at penetrations of 2.5 mm and 5.0 mm where the standard loads for the compacted crushed rock are 13.24 kN and 19.96 kN respectively.

Table D5.1 Load-penetration data for compacted crushed rock with CBR = 100 (BS1377 1990, Head 1988)

Penetration (mm)	2	2.5	4	5	6	8	10	12
Load (kN)	11.5	13.24	17.6	19.96	22.2	26.3	30.3	33.5

CBR is defined as:

$$CBR = \frac{\text{Plunger load on the specimen}}{\text{Plunger load for compacted crushed rock}} \times 100 \qquad (D5.1)$$

The higher of the two values computed at 2.5 mm and 5.0 mm is taken as the CBR. The load-penetration plots for different CBR values are shown in Figure D5.1. CBR values can be as low as 1 to 5 for fine grained soils and 10 to 60 for coarse grained soils. Compacted well graded gravels and the base courses of roads often have even higher CBR.

CBR tests in the laboratory are carried out in a 152 mm diameter CBR mold on intact or remolded specimens. The intact specimens can be taken from natural ground or compacted soils that are part of an embankment or roadwork. The remolded specimens can be compacted

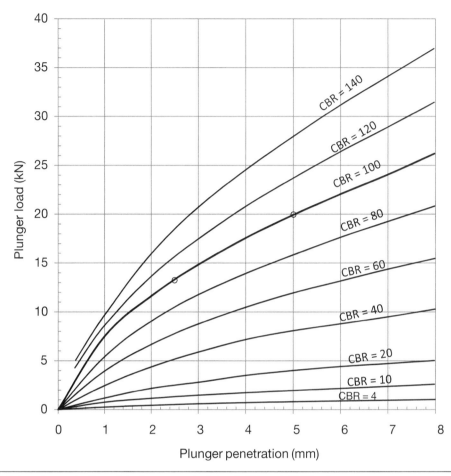

Figure D5.1 Load-penetration plots for geomaterials of different CBR values

by static or dynamic loads. They can also be tested after soaking in water for several days to simulate field situations. It is common to test the soils subjected to standard or modified Proctor compactive efforts at optimum water content or over a range of water contents. The test is generally carried out on a soil fraction finer than 19 mm.

Procedure:

Sample preparation:

1. Only a soil fraction finer than 19 mm should be used in the test. Any larger material should be replaced by 4.75-19 mm gravel fraction.
2. Carry out a standard or modified Proctor compaction test as appropriate (see compaction test), and determine the optimum water content and maximum dry density. This may be the basis for selecting the target water content and dry density of the CBR test specimen.

3. Mix the soil to a desired water content and cure for a minimum of two hours. For highly plastic clays a few days of curing is suggested.

4. Determine the mass m_1 of the mold. Figure D5.2 shows the CBR mold, perforated base plate, collar, and the spacer disc.

5. Place the 150 mm diameter and 61 mm thick spacer disc inside the CBR mold clamped to the perforated base plate and attach the collar as well. Place a filter paper on top of the spacer disk. The threaded handle is useful for lowering the spacer disk into the mold and for removal later.

6. Just before compaction, thoroughly mix the soil and determine the water content w_1 from a representative specimen.

7. Compact the soil in 3 to 5 equal layers as in the compaction test, using the appropriate number of blows.

8. Remove the collar and trim the surface of the compacted specimen level with the top of the mold, using a metal straightedge.

9. Remove the base plate, spacer disk, and filter paper and then determine the mass of the mold plus specimen m_2.

10. Place a filter paper on the perforated base plate and place the inverted mold on top with the compacted soil in contact with the filter paper at the bottom. This leaves a 61 mm gap at the top for accommodating the surcharge loads that simulate the confining pressure due to the overburden.

Figure D5.2 CBR mold, base plate, spacer disc, and collar (Photograph: N. Sivakugan)

Soaking (if required):

When it is required to carry out a soaked CBR test, the specimen should be soaked as follows:

1. Determine the mass of the base plate, mold, and specimen m_3.
2. Place the stem and perforated plate on the compacted soil specimen in the mold and apply a surcharge of at least 4.5 kg. The surcharge weights are in the form of annular or slotted 150 mm diameter disks.
3. Immerse the surcharged specimen in water for a period of 96 hours, allowing free access of water through the top and bottom of the specimen by maintaining the water level above the mold.
4. If swell measurements are required, use the tripod and dial indicator (see Figure D5.3) to measure the swell of the specimen during this period.
5. Remove the surcharges, stem, and the perforated plate as well as any surface water. Allow the specimen to drain downward for 15 minutes. Determine the mass of the base plate, mold, and specimen m_4.
6. Carry out the penetration test as soon as practicable before the specimen dries out.

Penetration Test:

The penetration test shall be performed on the end of the compacted specimen (soaked or unsoaked) that was in contact with the spacer disc during compaction. The procedure is as follows:

1. Prior to the seating of the plunger on the compacted soil surface, place the 2.25 kg annular surcharge on the soil surface and then place the mold, specimen, and base plate in the loading machine.
2. Seat the penetration piston with the smallest possible load; not exceeding 50 N for CBR < 30 and 250 N for CBR > 30.
3. Apply additional surcharge as required. Unless otherwise specified, the surcharge mass shall be 4.5 kg. If the specimen was soaked, apply surcharges equivalent in mass to those applied during soaking.
4. Initialize the force-measuring device and the displacement measuring device used to measure penetration. The penetration measured should be that of the piston relative to the mold.
5. Apply the load with a constant rate of penetration of 1 ± 0.2 mm/min. Record load readings at penetrations of 0.5, 1.0, 1.5, 2.0, 2.5, 3.0, 4.0, 5.0, 7.5, 10.0, and 12.5 mm (Figure D5.4).
6. Remove the soil from the mold and determine the moisture content of the top 30 mm layer w_{30} and, if required, that of the remaining specimen w_r.

Figure D5.3 Dial gage arrangement for measuring swell during soaking (Photograph: N. Sivakugan)

Figure D5.4 CBR penetration test (Photograph: N. Sivakugan)

Datasheet:

A simple datasheet for this test, prepared in Excel, is shown in Table D5.2.
The load-penetration plot is shown in Figure D5.5.

Table D5.2 CBR test datasheet

Sample description: Gravelly sandy clay of high plasticity

Sample location: Townsville port

Average specific gravity	2.80	Sample No.	TPA-43
Grading :		Date:	10th June 2010
< 4.75 mm fraction	78	Tested by:	Warren O'Donnell
4.75–9.5 mm fraction	15	Comments:	
9.5–19.0 mm fraction	7		

Compactive effort: Std. Proctor—3 layers, 60 blows/layer, 2.7 kg hammer

Molding water content (target):   17.5%

Bulk density ρ_m:	
Mold number	A6
Specimen volume, V_1 (cm³)	2123.0
Mass of mold, m_1 (g)	6802.0
Mass of mold + compacted soil, m_2 (g)	11228.9
Bulk density, ρ_m (Mg/m³)	2.09

Water content:	As compacted	Soaked Top 30 mm	Soaked Remainder
Tin number	42	45	61
Mass of tin (g)	31.67	73.67	82.13
Mass of tin + wet soil (g)	126.23	1053.40	3621.40
Mass of tin + dry soil (g)	112.32	882.30	3015.40
Water content (%)	w_1 17.25	21.16	20.66
Dry density, ρ_d (Mg/m³)	1.78 w_{30}		w_r
Mass of dry soil, m_s (g)	3775.7		

Soaking and swelling:	Notes:
Mass of mold + compacted soil, m_3 (g)	11228.9 Surcharge of 4.54 kg;
Mass of mold + compacted soaked soil, m_4 (g)	11276.4 soaked for 98 hours
Specimen height, h_i (mm)	117.0
Swell height, s (mm)	4.5
Swell percent, S (%)	3.8

Penetration test:						
Penetration (mm)	0	0.5	1	1.5	2	2.5
Prvng ring rdng	0	51	98	131	159	183
Load (kN)	0.00	0.64	1.23	1.64	1.99	2.29
Penetration (mm)	3	4	5	7.5	10	12.5
Prvng ring rdng	207	244	276	340	391	439
Load (kN)	2.56	3.02	3.41	4.20	4.83	5.43

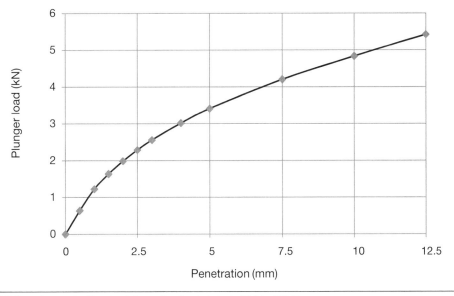

Figure D5.5 Load-penetration test from the test data in Table D5.2

At 2.5 mm and 5.0 mm penetrations, using Equation D5.1, CBR is the larger of:

$$\frac{2.29}{13.24} \times 100 = 17.3\%$$

and

$$\frac{3.41}{19.96} \times 100 = 17.1\%$$

In this case, both values are similar and CBR will be reported as 17. The CBR value reported should be rounded off as suggested in Table D5.3.

Table D5.3 Rounding off CBR when reporting (AS 1289.6.1.1)

CBR (%)	To the nearest
< 5	0.5
5 to 20	1
20 to 50	5
> 50	10

The other data recorded may include:

- Sample source and complete description
- Sample preparation and curing time

- Method of compaction and specimen preparation
- Dry density ρ_d and water content w_1 of the compacted specimen
- Swell percent and dry density after soaking ρ_{da} (if soaked)
- Water contents of the top 30 mm w_{30} and remainder w_r (if soaked)
- Surcharges applied
- CBR value and penetration at which CBR was determined
- Standard followed and deviations, if any

Analysis:

Load-penetration curve correction: Plot the load-penetration curve that can take one of the three forms shown in Figure D5.6. The convex upward curve *A* is the most typical, which requires no correction. When the load-penetration curve is concave upward initially, as in the case of curve *B*, due to surface irregularities or other reasons, adjust the zero point as shown in the figure. Here, the tangent is drawn at the inflection point. The intersection of the tangent and the penetration axis defines the new origin. In other words, the origin is shifted toward the right. If the correction is greater than 2 mm, the load-penetration curve shall be presented in the test report. In the case of curve *C*, where the penetration of the plunger increases the soil density and hence the strength, this correction is not applicable.

Read from the load-penetration curve, correct if required, the force value in kN at penetrations of 2.5 mm and 5.0 mm and calculate the bearing ratio for each by dividing by 13.2 kN and

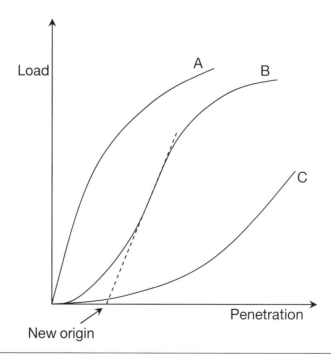

Figure D5.6 Load-penetration curve correction

19.8 kN, respectively, and multiplying by 100. Record the greater value of the calculated values as the CBR of the material.

Calculate the dry density of the specimen before soaking (ρ_d) from the following equation:

$$\rho_d = \frac{1}{V_1}\frac{m_2 - m_1}{\left(1 + \dfrac{w_1}{100}\right)}$$ (D5.1)

where

ρ_d = dry density of the specimen, in Mg/m^3
m_2 = mass of the mold (include base if necessary) plus compacted soil, in grams
m_1 = mass of the mold (include base if necessary), in grams
V_1 = volume of the specimen before soaking, in cubic centimeters (volume of the mold less the volume occupied by the disc)
w_1 = water content of the soil just before compaction, as a percent

The following computations are required only when the specimen is soaked prior to the penetration test.

If the swell is to be measured, calculate the swell percent (S) from the following equation:

$$S(\%) = \frac{s}{h_i} \times 100$$ (D5.2)

where

S = the swell of the specimen, as a percent
h_i = the initial height of the specimen (mm) before soaking
s = the change in the specimen height (mm)

Calculate the mass of dry soil in the specimen m_s from the following equation:

$$m_s = \frac{m_2 - m_1}{\left(1 + \dfrac{w_1}{100}\right)}$$ (D5.3)

If the specimen has been soaked, calculate the water content of the specimen after soaking (w_w) from the following equation:

$$w_w = w_1 + \frac{m_4 - m_3}{m_s} \times 100$$ (D5.4)

where

w_w = water content of the specimen after soaking, as a percent
w_1 = water content of the soil just before compaction, as a percent
m_4 = mass of mold, base plate, and specimen after soaking, in grams
m_3 = mass of mold, base plate, and specimen before soaking, in grams
m_s = mass of dry soil in the specimen, in grams

If the specimen has been soaked, calculate the volume of the specimen after soaking (V_2) from the following equation:

$$V_2 = V_1 \left(\frac{100 + S}{100} \right) \tag{D5.5}$$

where

V_2 = the volume of the specimen after soaking, in cubic centimeters
V_1 = volume of the specimen before soaking, in cubic centimeters (volume of the mold less the volume occupied by the disc)
S = the swell of the specimen, in percent

If the specimen has been soaked, calculate the specimen dry density after soaking ρ_{da} from the following equation:

$$\rho_{da} = \frac{m_s}{V_2} \tag{D5.6}$$

Cost: US$220 (one point) and US$120 (each additional point)

D6   Large Direct Shear Box Test

Objective: To determine the drained shear strength parameters of an aggregate through a direct shear test

Standards: ASTM D5321
AS 1289.6.2.2
BS 1377-7

Introduction

The large direct shear box test is carried out generally on soil samples containing large grains. The principles, test procedures, and the interpretations are similar to the standard direct shear tests (Test B19) carried out on smaller soil specimens.

Soils are generally modeled by Mohr-Coulomb constitutive model at failure, where the shear strength τ_f is related to the normal stress σ on the failure plane by:

$$\tau_f = c + \sigma \tan \phi \qquad (B19.1)$$

where c = cohesion and ϕ = friction angle of the soil. The equation defines the failure envelope in τ-σ plane. The shear strength parameters c and ϕ are different for drained and undrained loading situations. There is no provision to control the drainage during shearing in a standard direct shear apparatus. Therefore, it is generally recommended that direct shear test be carried out under drained conditions where the loading rate is slow enough to ensure that there is no buildup of excess pore water pressures. Nevertheless, it is possible to test the specimen under undrained conditions by shearing at a faster loading rate.

An oversimplified schematic diagram of a large direct shear test setup is shown in Figure D6.1 where the soil specimen is contained within a large, rigid metal box with square cross section in plan. The box is split horizontally with a tiny gap between the upper and lower sections. The soil specimen contained within the box is typically 305 mm in width and 204 mm in thickness. The bottom half is fixed to the frame of the apparatus whereas the top half can be moved relative to the top one with the help of an actuator controlled by an electric motor, thus shearing the soil specimen along the horizontal *failure plane* shown by the dashed line in Figure D6.1. Frequently, the upper box remains fixed while the lower box is mobile.

A normal load N is applied vertically which results in a normal stress σ applied within the specimen and on the failure plane. The normal stress σ is given by N/A where A is the cross-sectional area of the specimen. A shear load S is applied on the upper box and increased gradually from zero. The shear stress τ on the failure plane is given by S/A. One linear variable differential transducer (LVDT) is attached to the upper half of the apparatus to measure the vertical displacement (δ_v) of the specimen and another LVDT is attached to the bottom

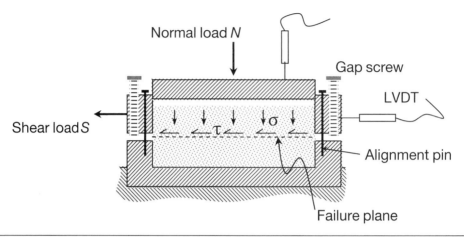

Figure D6.1 Large direct shear box test–schematic diagram

half of the apparatus to measure the horizontal displacement (δ_h) of the specimen. By plotting the shear stress τ against the horizontal displacement δ_h, the *shear stress at failure* (i.e., *shear strength*) τ_f can be identified. This procedure can be repeated for three different values of normal stresses. The corresponding values of τ_f can be plotted against σ, thus defining the failure envelope (Equation B19.1), from which c and ϕ can be determined.

Procedure:

1. *Soil specimen*: The test can be carried out on intact, remolded, or compacted specimens.
2. *Shear box*: Made of stainless steel with provisions for drainage from the bottom. The box is split along a horizontal plane into an upper and lower half, separated by a very thin gap and fitted together with two vertical alignment pins. Four gap screws are used to control the gap between the two halves.
3. *Perforations*: Provided at the bottom of the specimen to facilitate drainage.
4. *LVDTs*: Used for measuring horizontal and vertical displacements. Horizontal ones may have a stroke length of 150 mm and precision of ± 0.001 mm. Vertical ones will have the same precision of ± 0.001 mm.
5. *Shear load measuring device*: A load cell accurate to 0.001 N is for measuring the shear load S.
6. Assemble the shear box, keeping the two alignment pins in place to ensure that the top and bottom halves are aligned properly. Ensure that the gap screws are wound back so that they don't protrude and that the gap between the two halves is minimal. Place/compact the specimen in layers of known mass (m_1) and water content (w_1) sandwiched between the bottom of the lower half and the loading platen at the top. Determine the specimen dimensions. Attach the vertical LVDT.

7. Apply the desired normal load N and allow complete consolidation before applying the shear load. This can be ensured by recording the specimen compression against the time and following Casagrande's or Taylor's procedures discussed in the consolidation test section.
8. Remove the vertical alignment pins and use the four gap screws to open the gap between the two halves to 0.64 mm.
9. *Shearing*: Shearing should be carried out slowly such that there is no buildup of excess pore water pressure.
10. This apparatus is coupled with a data acquisition system. During consolidation and shearing, the following readings are taken: normal load N, shear load S, shear displacement δ_h, and vertical displacement δ_v. Continue shearing well beyond failure.
11. Repeat the above steps for three or more normal loads.

Notes:

1. For fine grained soils, the container holding the shear box can be filled with water for the entire duration of the test.
2. The direct shear test is generally strain-controlled rather than stress-controlled. The advantage of the strain-controlled test is that the peak and residual states can be defined more precisely in the case of a strain-softening soil.

Datasheet:

A simple datasheet for a direct shear test on an aggregate is given in Table D6.1.
The following information can be included in the test report:

- Soil classification, location, and project
- Type of shear device used
- Initial and final water contents
- Initial and final dry densities
- * Loading (strain) rate and shearing time for each load
- Plots of τ versus δ_h and δ_v versus δ_h for each specimen
- Ultimate and peak (if present) shear stresses for each normal stress and the two failure envelopes.

Analysis:

The initial dry density can be computed as:

$$\rho_d = \frac{m_1}{1 + w_1} \tag{D6.1}$$

Table D6.1 Datasheet for large direct shear box test on a recycled aggregate (Courtesy of Dr. Jegatheesan Piratheepan of Swinburne University of Technology)

Soil description	Recycled Asphalt				
Sample No.	RAP-50				
Location	Swinburne University of Technology				
Date	10-Oct-10				
Tested by	DIV				

Specimen mass, m_1 (g)	39700		Initial water content, w_1 (%)	8.1
Spec. gravity of the grains	2.68			
Specimen dimensions (mm):				
Length	305	Width	305	Thickness	194
Area (cm^2)	930.25		Volume (cm^3)	18047
Normal load (N)			Normal stress σ (kPa)	50.0

Time	δ_h (mm)	δ_v (mm)	S (N)	τ (kPa)
0	0	0	0.00	0.00
1	0.05132	0.0016	51.41	0.55
2	0.09622	0.0016	100.76	1.08
3	0.13631	0.0032	150.12	1.61
4	0.19565	0.0032	197.41	2.12
5	0.22452	0.0032	242.65	2.61
10	0.4763	0.0048	448.30	4.82
15	0.7168	0.0048	639.54	6.87
20	0.9558	−0.0016	810.23	8.71
25	1.1947	−0.0096	970.63	10.43
30	1.4369	−0.0224	1120.71	12.05
35	1.6791	−0.0352	1256.51	13.51
40	1.9212	−0.048	1400.41	15.05
45	2.1746	−0.0576	1544.41	16.60
50	2.4135	−0.0688	1657.51	17.82
60	2.9027	−0.0913	1889.81	20.32
70	3.379	−0.1585	2083.11	22.39
80	3.8745	−0.1793	2288.81	24.60
90	4.3636	−0.2305	2484.11	26.70
100	4.8431	−0.2897	2652.81	28.52
125	6.0972	−0.4257	3051.71	32.81
150	7.3	−0.5746	3452.71	37.12
175	8.5396	−0.7202	3790.01	40.74
200	9.7486	−0.8819	4047.01	43.50

Time	δ_h (mm)	δ_v (mm)	S (N)	τ (kPa)
225	10.9946	−1.0659	4260.91	45.80
250	12.2296	−1.2324	4454.21	47.88
275	13.4676	−1.4292	4610.51	49.56
300	14.7006	−1.6164	4733.91	50.89
325	15.9276	−1.8037	4859.31	52.24
350	17.1646	−2.015	4939.51	53.10
375	18.3906	−2.223	5027.91	54.05
400	19.6356	−2.39905	5176.01	55.64
425	20.8716	−2.5719	5213.01	56.04
450	22.1146	−2.76075	5235.61	56.28
475	23.3516	−2.93999	5223.31	56.15
500	24.5896	−3.10003	5293.21	56.90
550	27.0546	−3.372105	5305.61	57.03
600	29.5176	−3.64738	5301.41	56.99
650	32.0096	−3.91785	5219.21	56.11
700	34.4676	−4.14671	5231.51	56.24
750	36.9396	−4.3724	5122.51	55.07
800	39.4006	−4.5964	5085.51	54.67
850	41.8806	−4.7741	4974.51	53.48
900	44.3576	−4.9773	4768.81	51.26
950	46.8256	−5.1358	4836.71	51.99
1000	49.2956	−5.251	4838.71	52.02

where the initial water content is expressed as a decimal number (e.g., 0.155 instead of 15.5%). The final dry density can be computed in a similar manner from the final water content and the specimen dimensions.

The normal stress during shearing was computed as 50.0 kPa. As the specimen is sheared, with increasing horizontal displacement δ_h, there is a gradual reduction in the shearing area A' that may be computed as $A' = L(L\text{-}\delta_h)$. AS 1289.6.2.2 recommends using this reduced area in computing the shear and normal stresses. This correction is not recommended in ASTM D5321 and was not carried out for the data presented in Table D6.1 and Figure D6.2. The plots of τ and δ_v against δ_h are shown in Figure D6.2. The peak and ultimate shear strengths are identified as 57.0 kPa and 52.0 kPa, which are not very different. In case of dense sands and overconsolidated clays, there can be a significant drop in shear strength from peak to ultimate states. In normally consolidated clays and loose sands they are about the same, where the peak occurs at the ultimate state. Figure D6.3 shows the large direct shear box apparatus.

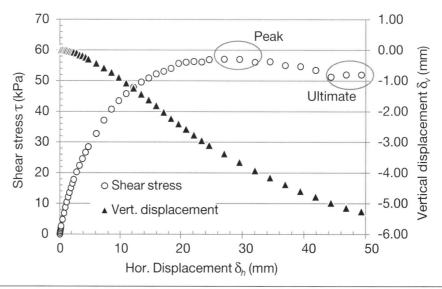

Figure D6.2 Plots of τ and δ_v against δ_h for data in Table D6.1

Figure D6.3 Large shear box (Photograph: N. Sivakugan)

By plotting the shear strength against the normal stress, from the data for three or more normal loads, the failure envelopes for peak and residual states can be defined, and, from these, the values of c and ϕ can be determined. In terms of effective stresses, ϕ_{peak} is greater than $\phi_{ultimate}$, and $c_{ult} \approx 0$.

Cost: US$2000–US$3000

D7   Los Angeles Abrasion Loss

Objective: To determine the loss, on abrasion, of aggregate particles by means of the Los Angeles (LA) abrasion testing machine

Standards: AS 1141.23*
ASTM C131 & ASTM C535

Introduction

This test is a measure of degradation of aggregates of standard gradings resulting from a combination of actions including abrasion or attrition, impact, and grinding in a 711 mm diameter and 508 mm long steel drum rotating about its horizontal axis (Figure D7.1). The test is carried out on aggregate fraction coarser than 1.70 mm (No. 12 sieve). The sample and a specific number of metal spheres (known as *charge*; see Figure D7.2) are tumbled inside the drum that rotates at 30 to 33 rpm for 500 rotations. The spheres are typically 46.8 mm in diameter with a

Figure D7.1 LA abrasion test machine
(Photograph: N. Sivakugan)

Figure D7.2 LA abrasion steel ball charges
(Photograph: N. Sivakugan)

Figure D7.3 LA abrasion steel ball charges
with recycled asphalt samples after testing
(Photograph: N. Sivakugan)

mass of 390 to 445 g, and are made of steel or cast iron. As the drum rotates, a shelf plate picks up the sample and the steel spheres, carrying them around until they are dropped to the opposite side of the drum, creating an impact crushing effect. The removable steel shelf projects radially 89 mm into the cylinder and extends across the entire length. The contents then roll within the drum with an abrading and grinding action until the shelf plate picks up the sample and the steel spheres, and then the cycle is repeated. After the prescribed number of revolutions, the contents are removed from the drum (Figure D7.3) and the aggregate portion is sieved to measure the degradation as percent of loss, which is the *Los Angeles abrasion loss*. ASTM C131 and ASTM C535 cover the procedures for small-size (finer than 37.5 mm) and large-size (coarser than 19 mm) coarse aggregates, respectively.

Typical values of Los Angeles abrasion loss can vary from 10% for hard igneous rocks to as high as 60% for soft limestones or sandstones.

Procedure:

1. *Sample preparation:* The test portion consists of a combination of aggregate fraction obtained by sieving, washing, and riffling material to comply with one of the four gradings recommended in Table D7.1 by ASTM C131. Record the mass of the test portion m_t to the nearest gram.

Table D7.1 Four recommended gradings of the test samples (after ASTM C131)

Sieve size (mm)		Mass (g)			
Passing	Retained on	A	B	C	D
37.5	25.0	1250 ± 25			
25.0	19.0	1250 ± 25			
19.0	12.5	1250 ± 10	2500 ± 10		
12.5	9.5	1250 ± 10	2500 ± 10		
9.5	6.3			2500 ± 10	
6.3	4.75			2500 ± 10	
4.75	2.36				5000 ± 10
Total sample mass (g)		5000 ± 10	5000 ± 10	5000 ± 10	5000 ± 10

2. *Charge:* The charge consists of a specific number of steel or cast iron spheres that have 46.8 mm diameter and mass of 390 to 445 g each (Figure D7.2). The number of spheres to be used in the test depends on the grading of the test portion as recommended by ASTM C131 in Table D7.2.

Table D7.2 Recommended charge (after ASTM C131)

Grading (see Table D7.1)	No. of spheres	Total mass of the charge (g)
A	12	5000 ± 25
B	11	4584 ± 25
C	8	3330 ± 25
D	6	2500 ± 25

3. Place the test sample of 5000 g inside the drum, along with the necessary charge. Rotate the drum at a speed of 30 to 33 rpm for 500 revolutions.
4. Remove the material from the drum. Make a preliminary separation of the sample using a sieve coarser than 1.70 mm (No. 12). Sieve the finer portion on a 1.70 mm sieve;

wash the entire material coarser than 1.70 mm and oven dry the coarse fraction at 105 to 110°C and determine its mass m_s.

Note:

1. The gradings suggested by the Australian Standard AS 1141.23 are slightly different. It suggests 10000 g of sample with 5000 g charge and 1000 revolutions for large-size aggregates. In aggregates complying with grading type B (see Table D7.1) and containing flaky particles it recommends using flaky and nonflaky particles in the ratio of 1:3.

Datasheet:

A simple datasheet for this test, prepared in Excel, is shown in Table D7.3. The Los Angeles abrasion loss must be given to the nearest integer.

Table D7.3 Los Angeles abrasion loss datasheet

Sample no.	1	2	3	4	5	6	7
m_t	5000	5000	5000	5000	5000	5000	5000
m_w	3600	3590	3580	3570	3595	3640	3645
LA value	28	28	28	29	28	27	27

The other data recorded may include:

*   Sample source, type, and nominal maximum particle size
*   Grading designation (Table D7.1) used for the test
*   Standard followed and deviations, if any
*   Analysis:

The Los Angeles abrasion loss can be calculated from the following equation:

$$\text{Los Angeles abrasion loss } (\%) = \frac{m_t - m_w}{m_t} \times 100 \tag{D7.1}$$

where m_t = initial dry mass of the test sample before the abrasion, and m_w = final dry mass of the test sample retained on a 1.70 mm sieve.

This book has free material available for download from the
Web Added Value™ resource center at *www.jrosspub.com*

References

Part E

Australian Standards:

AS 1141.12–1996. Methods for sampling and testing aggregates. Method 12: Materials finer than 75 (m in aggregates (by washing).

AS 1141.15–1999. Methods for sampling and testing aggregates. Method 15: Flakiness index.

AS 1141.5–2000. Method for sampling and testing aggregates. Method 5: Particle density and water absorption of fine aggregate.

AS 1141.6.1–2000. Method for sampling and testing aggregates. Method 6.1: Particle density and water absorption of coarse aggregate—weighing-in-water method.

AS 1141.23–2009. Methods for sampling and testing aggregates. Method 23: Los Angeles value.

AS 1289.2.1.1–2005. Methods of testing soils for engineering purposes. Method 2.1.1: Soil moisture content test—determination of the moisture content of a soil-oven-drying method (standard method).

AS 1289.2.1.4–2005. Methods of testing soils for engineering purposes. Method 2.1.4: Soil moisture content tests—determination of the moisture content of a soil—microwave oven-drying method (subsidiary method).

AS 1289.3.1.1–2009. Methods of testing soils for engineering purposes. Method 3.1.1: Soil classification tests—determination of the liquid limit of a soil—four point Casagrande method.

AS 1289.3.4.1–2008. Methods of testing soils for engineering purposes. Method 3.4.1: Soil classification tests—determination of the linear shrinkage of a soil—standard method.

AS 1289.3.5.1–2006. Methods of testing soils for engineering purposes. Method 3.5.1: Soil classification tests—determination of the soil particle density of a soil—standard method.

AS 1289.3.6.1–2009. Methods of testing soils for engineering purposes. Method 3.6.1: Soil classification tests—determination of the particle size distribution of a soil—standard method of analysis by sieving.

AS 1289.3.6.3–2003. Methods of testing soils for engineering purposes. Method 3.6.3: Soil classification tests—determination of the particle size distribution of a soil—standard method of fine analysis using a hydrometer.

AS 1289.3.9.1–2002. Methods of testing soils for engineering purposes. Method 3.9.1: Soil classification tests—determination of the cone liquid limit of a soil.

AS 1289.4.3.1–1997. Methods of testing soils for engineering purposes. Method 4.3.1: Soil chemical tests—determination of the pH value of a soil—electrometric method.

AS 1289.5.1.1–2003. Methods of testing soils for engineering purposes. Method 5.1.1: Soil compaction and density tests—determination of the dry density/moisture content relation of a soil using standard compactive effort.

AS 1289.5.2.1–2003. Methods of testing soils for engineering purposes. Method 5.2.1: Soil compaction and density tests—determination of the dry density/moisture content relation of a soil using modified compactive effort.

AS 1289.5.3.1–2004. Methods of testing soils for engineering purposes. Method 5.3.1: Soils compaction and density tests—determination of the field density of a soil—sand replacement method using a sand cone apparatus.

AS 1289.5.3.2–2004. Methods of testing soils for engineering purposes. Method 5.3.2: Soil compaction and density tests—determination of the field density of a soil—sand replacement method using a sand pouring can, with or without a volume displacer.

AS 1289.5.5.1 –2003. Methods of testing soils for engineering purposes. Method 5.5.1: Soil compaction and density tests—determination of the minimum and maximum dry density of a cohesionless material—standard method.

AS 1289.6.1.1–1998. Methods of testing soils for engineering purposes. Method 6.1.1: Soil strength and consolidation tests—determination of the California Bearing Ratio—standard laboratory method for a remoulded specimen.

AS 1289.6.2.2–1998. Methods of testing soils for engineering purposes. Method 6.2.2: Soil strength and consolidation tests—determination of the shear strength of a soil—direct shear test using a shear box.

AS 1289.6.4.1–1998. Methods of testing soils for engineering purposes. Method 6.4.2: Soil strength and consolidation tests—determination of the compressive strength of a soil—compressive strength of a saturated specimen tested in undrained triaxial compression without measurement of pore water pressure.

AS 1289.6.4.2–1998. Methods of testing soils for engineering purposes. Method 6.4.2: Soil strength and consolidation tests—determination of the compressive strength of a soil—compressive strength of a saturated specimen tested in undrained triaxial compression with measurement of pore water pressure.

AS 1289.6.6.1–1998. Methods of testing soils for engineering purposes. Method 6.6.1: Soil strength and consolidation tests—determination of the one-dimensional consolidation properties of soil—standard method.

AS 1289.6.7.1–2001. Methods of testing soils for engineering purposes. Method 6.7.1: Soil strength and consolidation tests—determination of permeability of a soil—constant head method for a remoulded specimen.

AS 1289.6.7.2–2001. Methods of testing soils for engineering purposes. Method 6.7.2: Soil strength and consolidation tests—determination of permeability of a soil—falling head method for a remoulded specimen.

AS 1289.6.7.3–1999. Methods of testing soils for engineering purposes. Method 6.7.3: Soil strength and consolidation tests—determination of permeability of a soil—constant head method using a flexible wall permeameter.

AS 1726–1993. *Geotechnical Site Investigations*.

AS 4133.4.1–2007. Method 4.1: Rock strength tests—determination of point load strength index. Methods of testing rocks for engineering purposes.

AS 4133.2.1.1–2005. Method 2.1.2: Rock porosity and density tests—determination of rock porosity and dry density—saturation and buoyancy techniques. Methods of testing rocks for engineering purposes.

AS 4133.2.1.2–2005. Method 2.1.1: Rock porosity and density tests—determination of rock porosity and dry density—saturation and calliper techniques. Methods of testing rocks for engineering purposes.

AS 4133.4.2.1–2007. Method 4.2.1: Rock strength tests—determination of uniaxial compressive strength of 50 MPa and greater. Methods of testing rocks for engineering purposes.

AS 4133.4.3.1–2009. Method 4.3.1: Rock strength tests—determination of deformability of rock materials in uniaxial compression—strength of 50 MPa and greater. Methods of testing rocks for engineering purposes.

AS 4133.3.4–2005. Method 3.4: Rock swelling and slake durability tests—determination of the slake durability index of rock samples.

AS 4439.1–1999. Wastes, sediments, and contaminated soils, Part 1: Preparation of leachates—preliminary assessment.

AS 4439.2–1997. Wastes, sediments, and contaminated soils, Part 1: Preparation of leachates—zero headspace procedure.

AS 4439.3–1997. Wastes, sediments, and contaminated soils, Part 1: Preparation of leachates—bottle leaching procedures.

ASTM International Standards:

ASTM C127–07. Standard method for density, relative density (specific gravity), and absorption of coarse aggregate.

ASTM C128–07a. Standard method for density, relative density (specific gravity), and absorption of fine aggregate.

ASTM C131–06. Standard test method for resistance to degradation of small-size coarse aggregate by abrasion and impact in Los Angeles machine.

ASTM C535–09. Standard test method for resistance to degradation of large-size coarse aggregate by abrasion and impact in Los Angeles machine.

ASTM D422–63(2007). Standard test method for particle-size analysis of soils.

ASTM D698–07. Standard test method for laboratory compaction characteristics of soil using standard effort (12400 ft-lbf/ft^3 (600 kN-m/m^3)).

ASTM D1557–09. Standard test method for laboratory compaction characteristics of soil using modified effort (56,000 ft-lbf/ft^3 (2,700 kN-m/m^3)).

ASTM D854–06e1. Standard test methods for specific gravity of soil solids by water pycnometer.

ASTM D1556–07. Standard test method for density and unit weight of soil in place by sand-cone method.

ASTM D1587–08. Standard practice for thin-walled tube sampling of soils for geotechnical purposes.

ASTM D1883–07. Standard test method for CBR (California Bearing Ratio) of laboratory compacted soils.

ASTM D2166–06. Standard test method for unconfined compressive strength of cohesive soil.

ASTM D2167–08. Standard test method for density and unit weight of soil in place by the rubber balloon method.

ASTM D2216–05. Standard test methods for laboratory determination of water (moisture) content of soil and rock by mass.

ASTM D2434–68. Standard test method for permeability of granular soils (constant head).

ASTM D2435–04. Standard methods for one-dimensional consolidation properties of soils using incremental loading.

ASTM D2487–06e1. Standard practice for classification of soils for engineering purposes (Unified Soil Classification System).

ASTM D2488–09a. Standard practice for description and identification of soils (visual-manual procedure).

ASTM D2850–03a. Standard test method for unconsolidated-undrained triaxial compression test on cohesive soils.

ASTM D2936–08. Standard test method for direct tensile strength of intact rock core specimens.

ASTM D2937–04. Standard test method for density of soil in place by the drive-cylinder method

ASTM D2974–07a. Standard test methods for moisture, ash and organic matter of peat and other organic soils.

ASTM D3080–04. Standard test method for direct shear tests on soils under consolidated drained conditions.

ASTM D3967– 08. Standard test method for splitting tensile strength of intact rock core specimens.

ASTM D4253–00. Standard test method for maximum index density and unit weight of soils using a vibrating table.

ASTM D4254–00. Standard test method for minimum index density and unit weight of soils and calculation of relative density.

ASTM D4318–10. Standard test methods for liquid limit, plastic limit and plasticity index of soils.

ASTM D4643–08. Standard test methods for determination of water (moisture) content of soil by microwave oven heating.

ASTM D4644–08. Standard test method for slake durability of shales and similar weak rocks.

ASTM D4767–04. Standard test method for consolidated undrained triaxial compression test for cohesive soils.

ASTM D4972–01. Standard test methods for pH of soils.

ASTM D5030–04. Standard test method for density of soil and rock in place by water replacement method in a test pit.

ASTM D5731–08. Standard test method for determination of the point load strength index and application to rock strength classifications.

ASTM D5873–05. Standard test method for determination of rock hardness by rebound hammer method.

ASTM D5084–03. Standard test methods for measurement of hydraulic conductivity of saturated porous materials using a flexible wall permeameter.

ASTM D6913–04. Standard test method for particle size distribution (gradation) of soil using sieve analysis.

ASTM D7012–07e1. Standard test method for compressive strength and elastic moduli of intact rock core specimens under varying states of stress and temperatures.

British Standards:

BS 1377 (1990). Methods of test for soils for civil engineering purposes. British Standard Institution, London.

BS 812–105.1:1989. Testing aggregates. Methods for determination of particle shape. Flakiness index.

BS 812–112:1990. Testing aggregates. Method for determination of aggregate impact value.

International Society of Rock Mechanics Suggested Methods:

ISRM (1978a). Commission on Standardisation of Laboratory and Field Tests. Suggested methods for determining tensile strength of rock materials, *International Journal of Rock Mechanics and Mining Sciences and Geomechanics Abstract*, 15(3), 99–103.

ISRM (1978b). Commission on Standardisation of Laboratory and Field Tests. Suggested methods for the quantitative description of discontinuities in rock masses, *International Journal of Rock Mechanics and Mining Sciences and Geomechanics Abstract*, 15(6), 319–368.

ISRM (1978c). Commission on Standardisation of Laboratory and Field Tests. Suggested methods for determining hardness and abrasiveness of rocks, *International Journal of Rock Mechanics and Mining Sciences and Geomechanics Abstract*, 15(3), 89–97.

ISRM (1979a). Commission on Standardisation of Laboratory and Field Tests. Suggested methods for determination of the uniaxial compressive strength of rock materials, *International Journal of Rock Mechanics and Mining Sciences and Geomechanics Abstract*, 16(2), 135–140.

ISRM (1979b). Commission on Standardisation of Laboratory and Field Tests. Suggested methods for determining water content, porosity, density, absorption and related properties and swelling and slake durability index properties. *International Journal of Rock Mechanics and Mining Sciences and Geomechanics Abstracts*, 16 (2), 143–156.

ISRM (1983). Commission on Testing Methods. Suggested method for determining the strength of rock materials in triaxial compression, *International Journal of Rock Mechanics and Mining Sciences and Geomechanics Abstract*, 20(6), 285–290.

ISRM (1985). Commission on Testing Methods. Suggested method for determining point load strength, *International Journal of Rock Mechanics and Mining Sciences and Geomechanics Abstract*, 22(2), 51–60.

Other References:

Aydin, A. (2009). "ISRM suggested method for determination of the Schmidt hammer rebound hardness: Revised version." *International Journal of Rock Mechanics and Mining Sciences*, 46(3), 627–634.

Bieniawski, Z. T. (1975). "The point load test in geotechnical practice." *Engineering Geology*, 9(1), 1–11.

Bishop, A. W., and Henkel, D. J. (1962). *The Measurement of Soil Properties in the Triaxial Test*, 2nd ed. London: Edward Arnold, 228 pp.

Bowles, J. E. (1986). *Engineering Properties of Soils and their Measurement*, 3rd ed. New York: McGraw-Hill.

Broch, E., and Franklin, J. A. (1972). "The point load strength test." *International Journal of Rock Mechanics and Mining Sciences*, 9(6), 669–697.

Brown, E. T. (ed.) (1981). *Rock Characterization Testing and Monitoring—ISRM Suggested Methods*. Oxford, UK: Pergamon Press, 211 pp.

Burmister, D. M. (1949). "Principles and techniques of soil identification." *Proceedings of Annual Highway Research Board Meeting*, National Research Council, Washington, D.C., 29, 402–433.

Casagrande, A. (1932). "Research on the Atterberg limits of soils." *Public Roads*, 13(8), 121–136.

Casagrande, A. (1936). "The determination of pre-consolidation load and its practical significance." Discussion D-34, *Proceedings of the First International Conference on Soil Mechanics and Foundation Engineering*, Cambridge, III, 60–64.

Casagrande, A. (1938). "Notes on soil mechanics–first semester." Harvard University (unpublished), 129 pp.

Das, B. M. (2008). *Soil Mechanics Laboratory Manual*. New York: Oxford University Press, 432 pp.

Day, R. W. (2001). *Soil Testing Manual*. New York: McGraw-Hill, 618 pp.

Deere, D. U. (1964). "Technical description of rock cores for engineering purposes." *Rock Mechanics and Engineering Geology*, 1, 17–22.

Franklin, J. A., and Chandra, J. A. (1972). "The slake durability test." *International Journal of Rock mechanics and Mining Sciences*, 9(3), 325–341.

Gamble, J. C. (1971). *Durability–Plasticity classification of shales and other argillaceous rocks*. PhD Thesis, University of Illinois, 159 pp.

Germaine, J. T., and Germaine, A. V. (2009). *Geotechnical Laboratory Measurements for Engineers*. Hoboken, NJ: John Wiley & Sons, 351 pp.

Gercek, H. (2007). "Poisson's ratio values for rocks." *International Journal of Rock Mechanics and Mining Sciences*, 44(1), 1–13.

Handin, J. (1966). "Strength and ductility," *Handbook of Physical Contacts*. ed. S. P. Clark, Geological Society of America, New York, 223–289.

Hansbo, S. (1957). "A new approach to the determination of shear strength of clay by the fall-cone test." *Royal Swedish Geotechnical Institute Proceedings*, Stockholm, 14, 5–47.

Harr, M. E. (1977). *Mechanics of Particulate Media*. New York: McGraw-Hill.

Hawkes, I., and Mellor, M. (1970). "Uniaxial testing in rock mechanics laboratories." *Engineering Geology*, 4(3), 179–285.

Hazen, A. (1911). Discussion of "Dams on sand foundations." by A. C. Koenig, *Transactions*, ASCE, 73, 199–203.

Head, K. H. (2006). *Manual of Soil Laboratory Testing*, 3rd ed., vol. 1: Soil classification and compaction tests. Caithness, SCT: Whittles Publishing, 416 pp.

Head, K. H. (1992a). *Manual of Soil Laboratory Testing*, 2nd ed., vol. 1: Soil classification and compaction tests. London: Pentech Press, 388 pp.

Head, K. H. (1992b). *Manual of Soil Laboratory Testing*, 2nd ed., vol. 2: Permeability, shear strength and compressibility tests. London: Pentech Press, 747 pp.

Head, K. H. (1992c). *Manual of Soil Laboratory Testing*, 2nd ed., vol. 3: Effective stress tests. London: Pentech Press, 1238 pp.

Hilf, J. W. (1975). "Compacted Fill," Chapter 7, *Foundation Engineering Handbook*, eds. H. F. Winterkorn and H-Y. Fang, New York: Van Nostrand Reinhold.

Hoek, E., and Brown, E. T. (1980). "Empirical strength criterion for rock masses." *Journal of Geotechnical Engineering Division*, ASCE, 106(9), 1013–1035.

Hoek, E., and Brown, E. T. (1980). *Underground Excavations in Rocks*, rev. 2nd ed. London: The Institution of Mining and Metallurgy, 527 pp.

Hoek, E., and Brown, E. T. (1997). "Practical estimates of rock mass strength." *International Journal of Rock Mechanics and Mining Sciences & Geomechanics Abstracts*, 34(8), 1165–1186.

Holtz, R. D., and Kovacs, W. D. (1981). *An Introduction to Geotechnical Engineering*. Englewood Cliffs, NJ: Prentice-Hall.

Hondros, G. (1959). "The evaluation of Poisson's ratio and the modulus of materials of a low tensile resistance by the Brazilian (indirect tensile) test with particular reference to concrete." *Australian Journal of Applied Science*, 10(3), 243–268.

Hvorslev, M. J. (1949). *Subsurface exploration and sampling of soils for civil engineering purposes*, Waterways Experiment Station, U.S. Army Corps of Engineers, Vicksburg, VA.

Lambe, T. W. (1951). *Soil Testing for Engineers*, New York: John Wiley & Sons.

Lambe, T. W., and Whitman, R. V. (1979). *Soil Mechanics, SI Version*. New York: John Wiley & Sons.

Leroueil, S., and Le Bihan, J-P. (1996). "Liquid limits and fall cones." *Canadian Geotechnical Journal*, 33(5), 793–798.

McCarthy, D. F. (2007). *Essentials of Soil Mechanics and Foundations*, 7th ed. Upper Saddle River, NJ: Pearson Prentice-Hall.

Mellor, M., and Hawkes, I. (1971). "Measurement of tensile strength by diametral compression of discs and annuli." *Engineering Geology*, 5(3), 173–225.

Muskat, M. (1937). *The Flow of Homogeneous Fluids through Porous Media*. New York: McGraw Hill Book Company. Reprinted by J. W. Edwards, Ann Arbor, MI, 1946.

Peck, R. B., Hanson, W. E., and Thornburn, T. H. (1974). *Foundation Engineering*, 2nd ed. New York: John Wiley & Sons.

Rankine, K. J., Sivakugan, N., and Cowling, R. (2006). "Emplaced geotechnical characteristics of hydraulic fills in a number of Australian mines." *Geotechnical and Geological Engineering*, 24(1), 1–14.

Russell, A. R., and Wood, D. M. (2009). "Point load tests and strength measurements for brittle spheres." *International Journal of Rock Mechanics and Mining Sciences*, 46(2), 272–280.

Schmertmann, J. H. (1955). "The undisturbed consolidation behaviour of clay," *Transactions*, ASCE, 120, 1201–1233.

Schmidt, E. (1951). "A non-destructive concrete tester." *Concrete*, 59(8), 34–35.

Terzaghi, K. (1925). *Erdbaumechanik auf bodenphysikalischer grundlage*, Franz Deuticke, Leipzig und Wein, 399 pp.

Terzaghi, K., and Peck, R. B. (1967). *Soil Mechanics in Engineering Practice*, 2nd ed. New York: John Wiley & Sons.

Timoshenko, S. (1934). *Theory of Elasticity*. New York: McGraw-Hill.

Ulusay, R., and Hudson, J. A. (eds.) (2007). *The Complete ISRM Suggested Methods for Rock Characterization, Testing and Monitoring: 1974–2006*, ISRM Turkish National Group, Ankara, Turkey, 628 pp.

USACE (1986). *Laboratory Soils Testing*, EM-1110-2-1906 Washington, DC.

US EPA (1991). *Toxicity Characteristic Leaching Procedures*. Method 1311.

Wilson, S. D. (1970). "Suggested method of test for moisture-density relations of soils using Harvard Compaction Apparatus," *Special Procedures for Testing Soil and Rock for Engineering Purposes*, ASTM STP 479, Philadelphia, PA.

Wyllie, D., and Mah, C. W. (2004). *Rock Slope Engineering*, 4th ed. London: Spon Press, 432 pp.

Index